Al-Anbar Awakening

Volume II

Iraqi Perspectives

From Insurgency to Counterinsurgency in Iraq, 2004-2009

Edited by
Colonel Gary W. Montgomery
Chief Warrant Officer-4 Timothy S. McWilliams
U.S. Marine Corps Reserve

Marine Corps University
United States Marine Corps
Quantico, Virginia
2009

Marine Corps University Press
3078 Upshur Avenue
Quantico, VA
22134

1st Printing, 2009

PCN 10600001200

Al-Anbar Awakening

Volume II
Iraqi Perspectives

From Insurgency to Counterinsurgency in Iraq 2004-2009

Edited by Colonel Gary W. Montgomery and
Chief Warrant Officer-4 Timothy S. McWilliams

Foreword

by Major General John F. Kelly

Words like "won" or "victory" really do not apply when speaking of counterinsurgency operations. Insurgencies grow from problems and discontent within a given society. Solve the problems, and the insurgency goes away, as opposed to being defeated. The difficulty is that a government is not always willing to address the root causes of the insurgency because it is often the government itself that the insurgents want to eliminate.

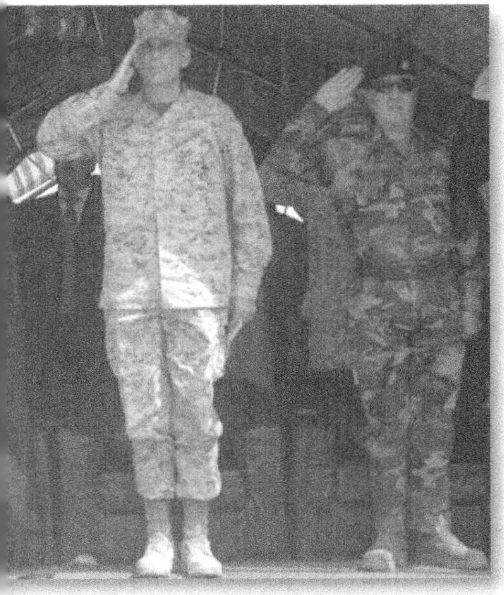

In Iraq to a very large degree, we—the U.S. military and civilians—were the source of the insurgency. Honest men and women can argue the whys, what-ifs, and what might-have-beens, but ultimately, it was mostly about unfulfilled promises and the heavy-handed military approach taken by some over the summer of 2003 that caused events to spiral out of control. No doubt the insurgency radicalized over time with al-Qaeda and Shi'a extremists playing a key role, but the insurgents did not initiate the war and only took advantage of the discontent.

If you asked Anbaris during my third tour in Iraq in 2008 why the insurgency began, most would look away and try to find a way not to answer. They would tell you that "we are friends now, and the causes are unimportant. It's all water under the bridge now." If pressed, they would talk about mutual misunderstandings and a lack of cultural awareness on both sides. They would say that expectations were too high on the part of the Iraqis about what America could do for them and how fast, but they seldom if ever blamed us directly. Press them further and they would mention the 29 April 2003 "massacre" in Fallujah, but more about the lack of an apology than the 70-plus unarmed citizens allegedly shot that day.

Another factor they would bring up was the shock and humiliation of having their army disbanded. The army was the one institution in Iraq everyone was proud of—Shi'a and Sunni alike—especially for what it had accomplished in protecting the nation against the Iranians in the 1980s. They perceived the disbanding as intentional contempt directed toward Iraq as a nation and as a people. They also saw it as the disarming of the nation. In the minds of many, this is when our status as liberators ended and that of occupier began.

Press the Anbaris one more time, and they would look you in the eye—but only if you are considered a friend—and they would state that after Baghdad fell and throughout the summer of 2003, the Americans overreacted to small acts of resistance or violence and fought in a way that was cowardly and without honor. Here they would talk about the senseless use of firepower and midnight raids on innocent men. They said that by our escalation, we proved true the rhetoric of the nationalist firebrands about why we had invaded, and our actions played directly into the hands of organizations like Zarqawi's al-Qaeda in Iraq and Sadr's militia.

Ask the same Anbari citizens why sometime in 2006 they began to turn against the by-then al-Qaeda-led insurgency, and the answer would be more direct. To them, their alliance with the radicals was a marriage of convenience to fight the U.S. occupation. Al-Qaeda brought dedication, organization, funding, and a willingness to die. Over time, however, it overplayed its hand and wore out its welcome by forcing an extreme Islamic agenda on a generally secular and very tribal culture. Al-Qaeda's campaign evolved from assistance, to persuasion, to intimidation, to murder in the most horrific ways, all designed to intimidate Anbari society—tribes and sheikhs alike—to adopt the most extreme form of Islam. At a certain point, al-Qaeda's agenda became too much for the average Anbari to bear. It was increasingly directed at the sheikhs themselves, and just as importantly, it began to have an impact on the business interests of tribal leaders.

The 17 paramount-dignified sheikhs of the major Anbari tribes and tribal federation turned away from al-Qaeda for survival purposes and toward U.S. forces for the same reason. They will tell you that Iraqis were being hunted down and killed by both the terrorists and the Coalition forces in Anbar. They knew the unbending terrorists would never meet them halfway, but they were confident that the Americans would—and they were right. Many of these men were once as much a part of the insurgency as Zarqawi was, albeit for different reasons. Over time, it became glaringly obvious to them that it was in their personal interests, and the interests of their tribes, to put a stop to the war.

When I returned to Iraq in February 2008 as commander of I Marine Expeditionary Force and Multi National Force-West (MNF-W), I was amazed at what I found. Violent incidents, once over 400 a week in al-Anbar Province, were down to 50 and had been in steady decline for months. Where Iraqis once avoided us, as any interaction jeopardized their lives and those of their families at the hands of al-Qaeda terrorists or nationalist insurgents, they were now aggressive in wanting to engage with us. Things had turned. The obvious questions were why had the change occurred, and was it sustainable, or was it simply due to an operational pause in the insurgent's effort? For months, Major General Walt Gaskin and his superb II Marine Expeditionary Force team, our immediate predecessors as MNF-W, had been wrestling with the answers. Their conclusions were ours to verify.

For MNF-W's part, since March 2004 we had extended the hand of friendship and cooperation, even as we were forced into a brutal fight that knew no quarter on the part of the Iraqi insurgents and foreign fighters. It was the major theme of our campaign plan, and it never changed. The command philosophy, a philosophy programmed into every Marine and U.S. Army unit that served in al-Anbar since we took the province, was that we had come to Iraq not to conquer, but to free, that we would always endeavor to "first, do not harm." This was often difficult, and sometimes you simply had to do a Fallujah II, even if Fallujah I had been ill-advised and totally counterproductive to what you were trying to do in the first place.

No single personality was the key in Anbar, no shiny new field manual the reason why, and no "surge" or single unit made it happen. It was a combination of many factors, not the least of which—perhaps the most important—was the consistent command philosophy that drove operations in Anbar from March 2004 forward. Each MNF-W commander and the troops under him continued to build upon the work of all those who came before. They took what their predecessors had done and ran with it, calling audibles as opportunities presented themselves. Consistency counts, and persistent presence on your feet puts you in more danger, no doubt, but also stacks the deck in your favor as you see more, hear more, know more, and engage more. It is these Americans—Marines, soldiers, sailors, and airmen, as well as civilians—who deserve the individual and collective credit for our part in the miracle that took place in al-Anbar Province. They slogged it out for more than six years to help the Anbaris create a miracle that spread to other regions of the country in late 2007, throughout 2008, and now into 2009.

I urge a note of caution to those who might have an overly inflated opinion of the role they played in the Awakening, or to the "experts" who write today as if they, with complete clairvoyance, predicted the change in loyalties in al-Anbar. The sheikhs, politicians, Iraqi security force officials, and even the former Ba'athist members of the military who reside in Anbar have a different opinion. They will tell you it was the sense of hopelessness the war had brought to the citizenry. The only hope for the future they could see was to be found in what members of MNF-W had done and were doing on their behalf despite the heat, the criticism from home, and the killing and casualties. They began to see us as a force that was sharing in their agony. Once they tried reaching out to some soldier or Marine's outstretched hand in friendship, it was over.

The interviews collected in the two volumes of this anthology do what no previous work has done—they attempt to tell the story of the al-Anbar Awakening from both sides, American and Iraqi. Not all the voices could be included, but there are many pertinent ones. The story they tell is a complex but important one, and it should be read with interest by all who want to truly understand what happened in Iraq between 2004 and 2009.

John F. Kelly
Major General, U.S. Marine Corps

Preface

This two-volume anthology of interviews tells the story of the al-Anbar Awakening and the emergence of al-Anbar Province from the throes of insurgency. It presents the perspectives of both Iraqis (volume two) and Americans (volume one) who ultimately came to work together, in an unlikely alliance of former adversaries, for the stabilization and redevelopment of the province. The collection begins in the 2003-2004 time frame with the rise of the insurgency and concludes with observations from the vantage point of early-to-mid 2009.

The anthology demonstrates that there is not one history of the Awakening, but several histories intertwined. It is not a complete collection, but one that provides a broad spectrum of candid, unvarnished perspectives from some of the leading players.

The American volume focuses on the roles and views of U.S. Marines, who were the primary Coalition force in al-Anbar from spring 2004 onward. At the time of their arrival, many military experts considered the province irredeemable. This collection chronicles the efforts of the Marines, and the soldiers, sailors, airmen, and civilians who worked with them, to consistently employ counterinsurgency tactics and to continue to reach out to the Iraqis during even the darkest days of the insurgency.

The Iraqi volume collects from many of the key Awakening players their views on how and why Anbaris came to turn against the insurgency that many had initially supported and seek the aid—both military and economic—of the Americans. Those interviewed include former Ba'ath Party military officers, senior officers in Iraq's new military, tribal sheikhs, Sunni imams, governmental representatives, and civilians.

This anthology is drawn from oral history interviews collected by field historians of the U.S. Marine Corps History Division, based at Marine Corps University in Quantico, Virginia. Field historians assigned to the History Division have collected hundreds of interviews since the beginning of Operation Iraqi Freedom I to serve as primary resources for future scholarship. In support of this anthology project,

Colonel Gary W. Montgomery and Chief Warrant Office-4 Timothy S. McWilliams deployed to Iraq in February and March of 2009 to interview Iraqis and additional American military and civilian personnel. Lieutenant Colonel Kurtis P. Wheeler had conducted more than 400 interviews in earlier deployments.

Like courtroom testimonies, oral histories are told from one person's perspective and may include discrepancies with, or even contradictions of, another witness's views. They are not a complete history, but they provide the outlines for one, to be fleshed out with documents and other sources not often collected or declassified this soon after events.

The interviews in this collection are edited excerpts drawn from longer interviews. They have been transcribed and edited according to scholarly standards to maintain the integrity of the interviews. Only interjections, false starts, and profanity have been silently omitted. Details added for clarity and accuracy are indicated by brackets. Omissions are noted by three-dot ellipses for partial sentences and four-dot ellipses for full sentences or more. With the Iraqi interviews, the interchange with interpreters has been omitted except in a few cases where the interpreter is attempting to clarify a point. Much of what has been left out of the American interviews is material that is duplicated in other interviews in the anthology. The full interviews and complete transcripts are part of the oral history collection of the Marine Corps History Division.

Ranks of officers, particularly American officers, reflect the rank at the time of the deployment under discussion. We have not tried to insert "then" in front of the ranks of all officers who have since been promoted.

We have attempted to verify the Iraqi person, place, and tribe names as best as possible, but undoubtedly there are several discrepancies, particularly in the Iraqi volume, where language barriers, dialects, the use of interpreters, and the mentions of many minor actors and areas made accurate transcription and identification challenging. There are also many variations in the transliteration of Iraqi names and terms.

* * *

The editors of this anthology acknowledge and thank a wide array of people for their support on this project. First and foremost, we thank the people whose stories are included for their time and candor. We particularly acknowledge Major General John F. Kelly, who wrote the foreword and who expedited the 2009 deployment of Colonel Montgomery and Chief Warrant Officer-4 McWilliams. In addition to the editors, those who conducted interviews included in the anthology are Colonel Jeffrey Acosta, Colonel Stephen E. Motsco, Colonel Michael D. Visconage, Lieutenant Colonel Craig H. Covert, Lieutenant Colonel John P. Piedmont, Lieutenant Colonel John R. Way, Staff Sergeant Bradford A. Wineman, Dr. David B. Crist, and Dr. Charles P. Neimeyer.

Dr. Neimeyer, director of the History Division; Mr. Charles D. Melson, chief historian; and Dr. Nathan S. Lowery, Field History branch head, provided guidance for the project. Mr. Kenneth H. Williams, senior editor for both the History Division and Marine Corps University Press, oversaw the editing and publication, assisted in the editing by Ms. Wanda J. Renfrow. Mr. Vincent J. Martinez provided layout and design for both volumes. Mr. Anthony R. Taglianetti, the History Division's oral historian, coordinated the timely transcription of the interviews. Lieutenant Colonel David A. Benhoff and Gunnery Sergeant Michael C. Coachman provided logistical support. Dr. Nicholas J. Schlosser, History Division historian, and Mr. Colin M. Colbourn, History Division intern, helped verify information.

Beyond the History Division, we are especially grateful to the interpreters. Those currently working in Iraq shall remain anonymous because of the inherent vulnerabilities peculiar to their vocation. Sometimes underappreciated and often overworked, their knowledge and perseverance was absolutely essential to our effort.

Many others labored to bring this project to fruition. Those who work outside of the normal publishing process are listed below. If we omitted anyone, it was inadvertent and not from lack of gratitude.

I Marine Expeditionary Force (Multi National Forces-West): Lieutenant Colonel Bradley E. Weisz (G-3 Air Officer); Lieutenant Colonel Todd W. Lyons (G-9 Foreign Affairs Officer/Marine Corps Intelligence Activity); Major Adam T. Strickland (Engagement Officer); 1st Lieutenant Timothy J.

Malham (Economic and Political Intelligence Center); Sergeant Luke O. Vancleave (Economic and Political Intelligence Center); Corporal Travis L. Helm (Economic and Political Intelligence Center); Corporal Lamont J. Lum (Economic and Political Intelligence Center); Lance Corporal Cassidy C. Niblett (Economic and Political Intelligence Center); Lance Corporal Orell D. Fisher (Economic and Political Intelligence Center).

II Marine Expeditionary Force: Colonel Robert W. Lanham (G-9 Assistant Chief of Staff); Lieutenant Colonel Bowen Richwine (G-9 Engagements OIC); Major Steven K. Barriger (G-9 Governance); 2d Lieutenant Anthony M. Bramante (Economic and Political Intelligence Center); Staff Sergeant William J. Rickards (G-9 Support); Sergeant Robert A. Pittenridge (G-9 Governance); Lance Corporal Thomas P. Wiltshire (Combat Camera); "Jack" Mahmood S. Al-Jumaily (Interpreter); Mythm Hassin (Interpreter).

Center for Advanced Operational Culture Learning, Quantico: Mr. Richard C. McPherson; Ms. Basema Maki (Interpreter); Mr. Hamid Lellou (Interpreter).

Marine Corps Intelligence Activity, Quantico: Colonel Philip D. Gentile (Commanding Officer); Mr. Dan J. Darling (Threat Analyst).

U.S. Marine Corps MARCENT LNO Cell, Kuwait: Gunnery Sergeant John M. Neatherton.

Marine Air-Ground Combat Training Center, Twentynine Palms, California: Staff Sergeant Michael A. Blaha (Combat Camera); Lance Corporal Ricky J. Holt (Combat Camera).

Marine Corps Base, Camp Pendleton, California: Captain Scott M. Clendaniel (Aide to General Kelly); Sergeant Eric L. Alabiso II (Combat Camera).

Introduction

I cannot describe the horror we lived in. Those were very bitter days. Those days we lived in hell. We looked like ghosts out of a cemetery. We were very tired. We had a lot of complaints to take to the Coalition forces, but we were afraid. Some days we wished to be dead just to be rested. But I have seven children. I was very tired.

Miriam, wife of a police officer,
describing life under al-Qaeda rule[1]

In an insurgency, the populace is the battlefield, and victory for the insurgent or the counterinsurgent tends toward the side better able to understand, influence, exploit, and satisfy the interests of the populace. Consequently, Miriam and hundreds of thousands of largely anonymous Anbaris were the final arbiters of the great issues of this period. So to understand the development of the insurgency, the rise of al-Qaeda in Iraq, and the al-Anbar Awakening, one must first understand the Anbaris.[2]

1. All block quotations in the introduction are from interviews in this volume. "Miriam" is a pseudonym for the first interviewee in the book.
2. This introduction is significantly informed by a massive report compiled by Marine Corps Intelligence Activity, "Study of the Insurgency in Anbar Province, Iraq," dated 13 June 2007.

The Anbaris

We are a country that is transforming from Bedouin to civilization. We depend on the elite and the notable. And these elite people are the link between us and you, the Americans. Where are the elite people? Where are the notable people? The civilization gap between us, I can say it reaches up to 200 years. Who is responsible for closing this gap between us? So you are the people who came from the future, and the elite people we're talking about understand you and understand Iraqi society.

General Ra'ad al-Hamdani
Former Republican Guard Corps Commander

General Ra'ad was referring to Iraq as a whole, but his comments are especially applicable to al-Anbar Province. Nearly all Anbaris are Sunni Arabs, and Sunni Arab elites have governed the Mesopotamian region for almost 500 years, first as surrogates under the Ottoman and British empires, and then outright in an independent Iraq. They had greater access to education, more opportunities for employment, and dominated the higher levels of government and the military. Over time, many developed a sense of entitlement and superiority relative to other Iraqis, and perhaps even a belief that they are the only ones capable of governing Iraq. If one's objective is a unified Iraq, then the latter opinion is also logical: the Kurds, though Sunnis, have strong separatist tendencies; and the Shi'a, though Arabs, are coreligionists with Iran, which is Iraq's arch enemy.

While al-Anbar's elites may be world-class in quality, they are relatively few in quantity. Their importance in society is probably reflected in the design of the first post-Ba'athist provincial council (roughly equivalent to a state legislature). The Anbaris chose to apportion representation according to vocational classification, not geographic precinct. Of the 49 seats comprising the council, 39 were designated for various professionals, such as doctors, engineers, and educators. The remaining 10 were reserved for tribal sheikhs.[3]

---

3. Interview with Mr. Kamis Ahmad Abban al-Alwani, deputy chairman of the provincial council, interview 11 in this volume.

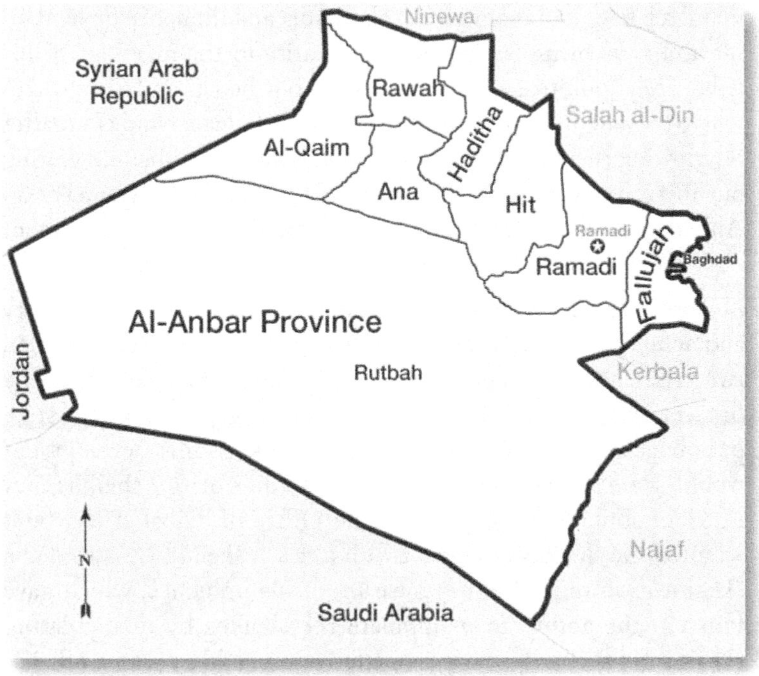

Districts of Al-Anbar Province

Al-Anbar is the most tribal of the 18 provinces of Iraq. The tribal system is poorly understood in the West, even though it is similar to the familiar, though defunct, Scottish clan system. In both systems, membership is based on kinship, loyalty is to clan or tribe ahead of the state, there are shifting rivalries and alliances among individuals as well as groups, and the leaders resist outside control, even while seeking support and patronage from the central government.

The two systems are also similar in that "sheikh" and "clan chief" are hereditary titles. However, this is where the systems diverge. The Anbari tribes do not practice primogeniture. A sheikh must be the son of a sheikh, but he may be any son of a sheikh. He is chosen by the consensus of the tribesmen; quite simply, he is the one whom the people trust and follow.

Furthermore, a sheikh's position is rarely secure. In this respect, he is similar to a Western politician. If his tribesmen (constituents) lose confidence in him, they will follow someone else—and he has many relatives and many rivals. His status and influence depend on his ability to bring patronage and security to the members of his tribe. They look to him for leadership, but he can only lead them where they are willing to go. Sometimes, tribal leadership is a matter of knowing the direction that the tribe is already going and getting out in front. Notably, in the interviews that follow, whenever an Anbari tribal leader made a significant decision, the announcement was preceded by an "education program" among his people.

Although the tribal system is deeply rooted in al-Anbar's history and actually precedes the introduction of Islam, in recent decades, two crises have endangered the system's continued existence. The first occurred when Saddam Hussein began providing state patronage to the tribesmen via their sheikhs. This development would seem to have enhanced the standing of the sheikhs, but Hussein simultaneously created the Office of Tribal Affairs and required the sheikhs to register with it. Each sheikh was assigned a classification reflecting the measure of his influence, which gave Hussein the ability to manipulate the sheikhs by manipulating classifications and patronage. In effect, the sheikhs became officials of the state. He also had his people register men without the requisite pedigree, thereby dramatically expanding the number of sheikhs. The result was the term "fake sheikh," as well as many tribal disputes that linger to this day.[4]

The second crisis occurred when al-Qaeda in Iraq began targeting tribal leaders, a tactic that intensified in late 2005. A number of sheikhs were assassinated, and many more either fled the country or were sent away by their tribes for their own protection.[5] Over time, the exiled sheikhs lost influence, while the younger men who remained behind in Iraq assumed many of their duties. Tribes were often (perhaps usually) internally divided in regard to which cause or faction to support during the insurgency.

For Americans serving in al-Anbar Province, there were additional complications. The title "sheikh" is also an honorific that is frequently

---

4. Lin Todd et al., "Iraq Tribal Study—Al-Anbar Governorate: The Albu Fahd Tribe, the Albu Mahal Tribe, and the Albu Issa Tribe" (Department of Defense study, 18 June 2006). See also the interview with Sheikh Majed Abd al-Razzaq Ali al-Sulayman in this volume.

5. See the interview with Col Michael M. Walker in the first volume of *Al-Anbar Awakening*.

used in addressing highly respected men or assumed by scoundrels seeking opportunities for personal gain at the expense of unwitting Americans. Also, a single sheikh may serve in many roles and represent different constituencies who have conflicting interests.[6]

Setting the Conditions for Insurgency

Everybody wanted to change the previous regime, but no one wanted it in this way. We really hoped that we could change the regime from inside Iraq, but it was too hard. No one could do it. Even all the other countries around the world accepted the idea that it wasn't going to happen from inside Iraq.

General Haqi Isma'eel Ali Hameed
Commander, 2d Region Directorate of Border Enforcement

Saddam Hussein led Iraq into one disaster after another. A year after assuming the presidency, he invaded Iran. The war dragged on from 1980 until 1988 with a devastating cost in blood and treasure. The economy was ruined and the national debt was impossible to meet. In 1990, just two years after the conclusion of the Iran-Iraq War, he invaded Kuwait, which resulted in the 1991 Gulf War. American-led Coalition air forces pummeled Iraqi infrastructure, military forces, and facilities for six weeks, and then Coalition ground forces drove the Iraqi army out of Kuwait in disorder.

Defeat opened the door for simultaneous insurrections in the Kurdish north and the Shi'a south. With 15 of 18 provinces in revolt, Hussein turned to the Sunnis for support. He warned that if the uprisings were successful, Iraq would become like Lebanon—a land of warring factions and utter chaos. Sunni fears of Iran and suspicion of Iraqi Shi'as were confirmed when Iraqi Shi'a forces of the Supreme Council for the Islamic Revolution in Iraq crossed the border from their safe haven in Iran to participate in the southern uprising. The Sunnis chose to support Saddam, and their response was swift and

---

6. For a detailed example, see William S. McCallister, "COIN and Irregular Warfare in a Tribal Society" (pamphlet, published by the *Small Wars Journal*, 2007), 30 (online at http://smallwarsjournal.com/documents/coinandiwinatribalsociety.pdf).

brutal. The largely Sunni Republican Guard crushed both uprisings with great slaughter, which included the use of chemical weapons in both the north and south.

Even though these actions left both Iraq and its leader weakened, Hussein refused to comply with the international economic sanctions. Instead, he tried to game his way out of them and failed, for more than a decade. In order to maintain his grip on power at home, he made major departures from Ba'ath Party ideology by embracing both the tribes and religion.

In Ba'ath socialism, the individual's duty to the state comes before his duty to his tribe. Therefore, tribalism was not only regarded by the Ba'athists as backward, but also as a threat to the government. Nevertheless, in shoring up his Sunni power base, Hussein granted greater autonomy and state patronage to the tribes. And, though Anbaris were not immune from the economic pain of international sanctions, Hussein ensured that they suffered less than most Iraqis. Government officials and military officers received extra rations, government cars, subsidized loans, and access to specialty shops with luxury goods. Furthermore, the sanctions increased the profits to be made from smuggling, which became a pillar of the Anbar economy. Anbaris used their transnational tribal connections to smuggle scrap metal, sheep, and oil into Jordan and Syria, where they were relatively expensive. They returned to Iraq with cigarettes, alcohol, and electrical appliances. These same smuggling routes and connections eventually served equally well as "ratlines" during the insurgency.

Regarding the second change, a religious revival transformed Iraq in the 1990s. Iraqi Shi'as had become more religious following the 1979 Iranian Revolution, and Iraqis of all sects and ethnicities had witnessed the religiously inspired enthusiasm of Iranian troops during the Iran-Iraq War. By the 1990s, religion was increasingly supplanting Ba'ath Party ideology, which had lost its relevance. Hussein disapproved of this trend, but unable to turn back the tide, he decided to co-opt it with a national "Return to Faith" campaign. Piety became a facet of Saddam's public image. The government closed bars and cinemas and spent large sums on mosques, clerics received salary increases and new services, and compulsory Quranic classes were instituted at all levels of education and even within the Ba'ath Party. Consequently, mosques became social centers, and clerical prestige increased until imams rivaled local sheikhs and party leaders in social influence. The secular Ba'ath Party was Islamicized, most people became more religious, and some became Islamic radicals.

As 2003 approached, Hussein prepared for the Gulf War of 1991. More specifically, he prepared for a repeat of the domestic uprisings that followed the Gulf War. Believing that the Coalition would be stopped by international political pressure and that Americans were too casualty-averse to conduct the ground operations necessary to overthrow his regime, he foresaw the worst-case scenario as a protracted air campaign followed by a limited invasion. Therefore, the bigger threats to his way of thinking were domestic uprisings and the potential for Iranian intervention.

So far, no evidence has surfaced suggesting that Hussein developed any plans for a post-invasion insurgency. Nevertheless, his war preparations had an irregular element, which served equally well for suppressing an uprising or starting an insurgency. Specifically, he established irregular and paramilitary organizations, such as the Saddam Fedayeen and al-Quds Army, which were tasked with containing and suppressing an uprising until the Republican Guard could arrive and destroy those who were rebelling. Second, he gave large quantities of weapons to loyal tribal sheikhs for the purpose of guarding the borders and suppressing dissidents. And third, in early 2003, Hussein decentralized command and control of the military in order to ensure that commanders had sufficient forces, support, and authority to respond quickly at the earliest signs of an uprising. He thereby enabled them to conduct preliminary planning for guerrilla and terrorist activities without central direction.

As war appeared imminent, some government mosques prepared the public for war by preaching anti-American and anti-Israeli sermons and urged Iraqis to jihad and resistance if the Americans invaded the country. This message was reinforced when the Islamic Research Center at al-Azhar University in Cairo, most prestigious Sunni theological center in the Islamic world, announced that an American invasion of Iraq would compel all Muslims to jihad. This pronouncement further legitimized anti-Coalition violence even among Islamists who hated the Saddam regime.[7]

Meanwhile, actions (or the absence thereof) by those aligning again Hussein also set conditions for insurgency. In January 2003, U.S. President George W. Bush established the Organization for Reconstruction and Humanitarian Assistance inside the Department of Defense to develop plans for democratic governance and rebuilding Iraq following regime change. The United States developed a branch plan for insurgency, but it did not write a detailed contingency plan, and it allocated no resources for counterinsurgency in the final war plan.

Development of the Insurgency

In March 2003, the Coalition invaded Iraq in Operation Iraqi Freedom. To everyone's surprise, Baghdad fell in a matter of weeks. The first Coalition forces to enter al-Anbar Province were British, Australian, and American special operations forces who were searching for Scud missiles in the western desert, hunting Ba'ath regime leaders, and securing important facilities and infrastructure such as Haditha Dam and al-Asad air base. They were followed by conventional forces from the U.S. Army. Since al-Anbar was only lightly defended, there were a couple of sharp engagements but no major battles.

In post-Saddam Hussein al-Anbar, resistance was initially disorganized and engaged in by small groups or individuals. Still, most Anbaris were cooperative with Coalition forces.

In May 2003, Ambassador L. Paul Bremer III announced the de-Ba'athification policy and the dissolution of the Iraqi Armed Forces, which threw hundreds of thousands of Iraq's most capable men into the ranks of the unemployed with no hope of future prospects.[8] A great many of these men were Anbaris.

---

7. Associated Press, 10 March 2003.
8. "De-Ba'athification of Iraqi Society" (Coalition Provisional Authority Order Number 1, 16 May 2003); "Dissolution of Entities" (Coalition Provisional Authority Order Number 2, 23 May 2003).

The number of armed groups proliferated and violence increased. These groups varied greatly in organization, capabilities, and goals, but they fell generally within certain broad categories. Sunni nationalists could be subdivided into two groups: former regime loyalists, who wanted the Ba'ath regime and Saddam Hussein restored to power; and former regime elements, who wanted the former regime restored without Hussein. There were also Sunni religious extemists, and another category, smaller than the others, of al-Qaeda-associated organizations.

The Coalition responded to the developing insurgency by focusing on tracking down former regime leaders. By the end of the year, Husay and Qusay Husseinwere dead and Saddam Hussein was in custody. The former regime loyalists became dispirited and splintered by rivalries. In defeating this segment of the insurgency, however, Coalition force methods contributed to the alienating the population and strengthened other elements of the insurgency.

Also in 2003, Abu Musab al-Zarqawi, an al-Qaeda associate and leader of Jama'at al-Tawhid wal-Jihad (Group of Monotheism and Jihad) moved his center of operations from the Kurdish north to al-Anbar, Diyala, and Baghdad. Over the next year, he kept a low operational profile while building his organization through the absorption of discouraged former regime loyalists and by attracting foreign fighters. In early 2004, he did not think that he was receiving sufficient support from the Iraqis, and in a communication with al-Qaeda senior leadership, he argued for provoking a sectarian war against the Iraqi Shi'as in order to radicalize the Iraqi Sunnis. The proposal was rejected because it would jeopardize Iranian support for al-Qaeda.

By October of 2004, al-Zarqawi agreed to subordinate himself to al-Qaeda's higher objectives, and he announced that Jama'at al-Tawhid wal-Jihad was incorporated into al-Qaeda. In return, he received access to al-Qaeda's global network of resources, personnel, propaganda support, and financing.

By early 2004, the former regime elements and the Sunni religious extremists had matured from splintered groups into organized forces that were able to engage in pitched battles. Incompatible goals prevented the unification of these two major categories of insurgents, or even the unification of the various factions within a single category. Nevertheless, the creation of local coordinating councils enabled short-term tactical cooperation.

Increasing cooperation brought about a rise in Islamic sentiment among the nationalists. The rank and file began to see Iraqi supporters of the Coalition as apostates as well as collaborators; the Coalition Provisional Authority was an enemy of Islam as well as an occupying power; and the Shi'a were heretics as well as agents of Iran. However, neither the nationalists nor the Sunni religious extremists were willing to support al-Qaeda in Iraq or endorse mass casualty attacks.

In April 2004, the insurgent groups felt strong enough to face U.S. Marines in open battle in Fallujah. They were no match for the Marines, but political pressure forced the Marines to halt their attack (Operation Vigilant Resolve) and withdraw from the city. The insurgents regarded this as a victory, and insurgent morale soared.

Subsequent negotiations between the Marines and insurgents resulted in the establishment of the Fallujah Brigade, which was intended to provide local control of security within Fallujah. However, the unit was compromised by insurgents and soon disbanded.

While events attendant to the Fallujah Brigade were unwinding in al-Anbar, Talal al-Gaoud was hosting meetings with U.S. Marine officers and officials of the U.S. Department of Defense in Amman, Jordan, in an attempt to negotiate an agreement between the Americans and Anbar's secular elites.[9] Talal al-Gaoud was a member of the al-Gaoud subtribe of the Albu Nimr tribe. The Albu Nimr rebelled against Saddam Hussein in the 1990s, and Talal al-Gaoud's relative, Fasal Raikan Najris al-Gaoud, was appointed governor of al-Anbar Province not long after the negotiations in Amman ended. While the negotiations did not result in an agreement, the meetings were an early indication that influential Anbaris were interested in engagement.

In November 2004, as Fasal al-Gaoud assumed the governorship, U.S. Marines and soldiers were storming Fallujah (Operation Phantom Fury/al-Fajr). The major insurgent groups suffered a severe blow. Thousands of fighters were killed or captured. The leadership spent several months reorganizing, and their men retreated into their remaining strongholds in western al-Anbar, unable to interfere with elections in January 2005.

However, antipathy for the Coalition led to a Sunni boycott of the election. Coalition diplomatic efforts during the remainder of the year focused on engaging Sunni leaders in the political process. These efforts, and a sense of hopelessness among insurgents following the storming of Fallujah, eventually bore some fruit. Sunni voter turnout

---

[9] David Rose, "Iraq: Heads in the Sand," *Vanity Fair*, 12 May 2009.

increased for the summer's constitutional referendum and the December 2005 election. These developments created a division between the more nationalist insurgents, led by Mohammed Mahmoud Latif, and the Islamists, led by Abu Musab al-Zarqawi.

The Rise of al-Qaeda in Iraq

Beginning in April 2005, insurgents stepped up attempts to target Baghdad, which led to Coalition military action against their strongholds in western al-Anbar. By the end of the year, insurgent participation in the political process was increasing, and nationalist motivation for the insurgency was declining. In-fighting ensued, which paved the way for the eventual dominance of al-Qaeda in Iraq the following year.

In late 2005, Operation Sayyad II had created a serious disruption in the al-Qaeda in Iraq network in western al-Anbar, which continued into 2006. Many other insurgents and emerging political groups used the opportunity to break away and seek political engagement with the faction under the direction of Mohammed Mahmoud Latif. Al-Qaeda in Iraq regrouped by early 2006, however, and it began targeting insurgent and tribal leaders who favored political engagement. This strategy was so successful that al-Qaeda in Iraq became the dominant insurgent group in all of al-Anbar. The magnitude of its success brought unparalleled power and created difficulties in maintaining the ideological integrity of the organization. Al-Qaida's murder and intimidation campaign also alienated large numbers of Anbaris, which created opportunities for both Coalition forces and insurgent rivals. The death of Abu Musab al-Zarqawi in June 2006 reinforced these opportunities, but it also allowed less volatile leaders to step forward within the organization, including Abu Ayyub al-Masri.

Al-Anbar Awakening

According to the Anbaris, the Albu Mahal tribe in al-Qaim initiated what became the Awakening in 2005 when they engaged in open warfare against al-Qaeda in Iraq throughout the spring and summer. This Awakening did not spread, probably because only local tribal interests were at stake. Nevertheless, tribes at the opposite end of al-Anbar were aware of it, and it seems to have served as a precedent.

The celebrated Awakening, which cleared the entire province, probably began in early 2006 and perhaps even in late 2005. It is difficult to determine a precise start date because the Al-Anbar Awakening began as the insurgency began: secretly and separately, with differing organizational structures, various capabilities, tactics adapted to local circumstances. Many of these stories and details can be found in the interviews that follow. By September 2006, the separate elements of the Awakening had coalesced and matured to the point where the movement was prepared to go public.

By the summer of 2007, al-Anbar Province was largely cleared of al-Qaeda in Iraq. Without the common bond of a common enemy, the Awakening began to splinter as the groups that had come together started to pursue divergent interests.

This is an outline of the backdrop against which the Awakening was set. The story of how it played out follows in the words of many of the brave Anbaris who made it happen.

* * *

In addition to the details in the preface on the editorial method applied across both volumes of this collection, here are some additional notes on the Iraqi interviews. The interviews in this volume were conducted between English and Arabic speakers who communicated through interpreters. Native Arabic speakers were integrally involved in each interview, using consecutive, and occasionally simultaneous, interpretation.

The single exception to these practices was the interview with the woman we has called Miriam (the first interview in the volume). The alternative procedures used to protect her identity are described in the introductory note to her interview.

Each interview was recorded on electronic recording devices. Upon the interviewers' return to Quantico, Virginia, the English portions were transcribed by a contracted transcription service. For publication, the transcripts were edited for clarity and abridged for relevance. Problematic passages were taken to the Marine Corps Center for Operational Culture and Learning (CAOCL), where native Arabic speakers listened to the original recordings and clarified the meanings. The resulting changes were reviewed and edited again, as necessary.

In the interviews, the Iraqis normally distinguished between Britons and Americans; however, they often used the terms soldier and Marine or Army and Marines interchangeably.

The Iraqis tend to use the word "division" in the same manner that English speakers use the word "unit." Therefore, only actual division-sized units are indicated by a capital letter.

The tribal system is hierarchical and contains words for confederations, tribes, clans, and families. As is the usual practice in interpreted conversation, tribe is most commonly used here.

Gary W. Montgomery
Colonel, U.S. Marine Corps Reserve

Iraqi Perspectives

A View of
Daily Life

10
CYALUME
LIGHTSTICKS

Interview 1

Miriam

Head of an Iraqi Women's Nongovernmental Organization

"Miriam" is a pseudonym. All personally identifiable information has been concealed for her safety and for the security of her family. Her husband is a high-ranking Iraqi policeman who was targeted by insurgents. Also, Miriam is prominent in her own right due to her humanitarian and political activities.

In the first half of the interview, she discusses daily life during the insurgency and her efforts to support her family while concealing her husband. In the second half, she describes her interaction with Marine civil affairs and the activities of her nongovernmental organization.

Miriam was interviewed by Colonel Gary W. Montgomery and Chief Warrant Officer-4 Timothy S. McWilliams on 17 February 2009 at the Provincial Government Center in Ramadi. In accordance with Miriam's wishes, audio and video equipment were not used. Colonel Montgomery, Chief Warrant Officer-4 McWilliams, and an Iraqi interpreter made handwritten notes, and the separate versions were merged into a single draft. Miriam reviewed the draft and made corrections. Miriam was accompanied by three Iraqi women who occasionally interjected additional details, but the words reproduced here are primarily Miriam's. Also present were Captain Stephanie [surname deleted], a Marine civil affairs officer who was on her fourth deployment to Iraq, and two female Iraqi interpreters, one translating and the other taking notes.

Miriam: At the beginning, the Iraqi police worked openly under Chief Ja'adan.* In 2003 to 2004, they were coming to work freely and working very well. When the terrorists started in al-Anbar, they attacked the police first, causing a massive scare. During the night, al-Qaeda threw threatening notes onto their doorsteps or into their gardens so that they would find them in the morning

* Major General Ja'adan Muhammad al-Alwani, Provincial Director of Police, 2003-2004.

before work. The notes told them to quit working for the police and to go to a mosque, publicly declare that they are no longer police, and ask for forgiveness.

The terrorists called for jihad to get public support, but they misused the word. Jihad is for defending our home and family against people trying to take us by force. It does not mean cutting people into little pieces. The terrorists used the word as a cover for their evil acts. They used the word to win the sympathy of the people. It is a very big word. The terrorists targeted illiterate minds—a certain level of society—and used them as a front.

They threatened the Iraqi police and made them quit. Then when the policemen were unemployed, they offered them a lot of money to work for al-Qaeda. People who didn't cooperate got "the discipline." Whoever did not cooperate was taken from his house in a hood, stripped down to his boxers, and whipped with a bike chain or cable in front of the people to set an example.

The ugliest torture was committed by al-Qaeda. If the discipline didn't work, the people were abducted and slaughtered. The head was put in a container and thrown away, or the neck cut and the head placed on the back.

In 2005, our 16-year-old neighbor was killed during Ramadan. At the beginning of Allahu Akhbar,* he was hanged from a three-story flag pole. When the last note sounded, his body was dropped.

My cousin's husband worked at a police station. They drove a tanker through the station and detonated it. Everyone was killed or burned. He lived, but he was badly burned. His wife took him to a doctor for treatment. The doctor referred him to Ramadi General Hospital. However, she was afraid to go there because al-Qaeda had taken over the hospital, so she treated him at home. Al-Qaeda suspected he was there, so they sent some ladies to the house to find out. A lot of women joined al-Qaeda. These women were seamstresses, and they managed to get in the house under the

---

* A broadcast from the mosque that signals that eating is permitted.

pretense of discussing some sewing work. They saw the husband and informed al-Qaeda. The house was raided and the husband abducted. They put rocks on his legs and put him into the river.

Al-Qaeda brainwashed people. If we heard gunshots, we went inside our homes. It was a signal that they were going to do something. They announced on a loudspeaker why they were killing.

Al-Qaeda also incited hatred between the Shi'a and the Sunni. They accused everyone wearing a precious stone on his ring of being a Shi'a and murdered him. Sometimes they fired shots into doors to scare them.

One time they caught three men going to Baghdad and killed them. They killed three young, tall men who were wearing rings with precious stones. They said that because they were wearing rings with precious stones that they were Shi'a. Our [Sunni] men don't usually wear rings with precious stones. They tortured, killed, and mutilated them and then left their bodies on the street for three days. They told everyone not to touch them. But their families recognized them. They were former policemen. They weren't Shi'as, they were our men.

They killed doctors and leaders. They killed doctors and said that it was because they treated Americans. The doctors fled the country. They killed mullahs and said that it was because they liked Americans. Soon there were no men left to kill, so they started killing women and children. They killed women and said that it was because their husbands were policemen. They killed children and said it was because their fathers were policemen.

I cannot describe the horror we lived in. Those were very bitter days. Those days we lived in hell. We looked like ghosts out of a cemetery. We were very tired. We had a lot of complaints to take to the Coalition forces, but we were afraid. Some days we wished to be dead just to be rested. But I have [many] children. I was very tired.

My husband let his beard and hair grow so that they wouldn't think he's working. We had no income, and I started selling our property to pay the rent. I heard that this officer was killed, that officer was killed. I sent my husband to hide sometimes.

Al-Qaeda started rumors. On the streets, they accused Americans of being killers. People welcomed al-Qaeda into their homes as a custom, but they killed families, raped wives, and took over. If one of them killed 60 people, he was called emir.

Interpreter: Each al-Qaeda cell consisted of six people. If you kill 60 people, then you become an emir automatically. The cells didn't know each other.

Miriam: Each home they take over becomes a cell that doesn't work with the others. Every group of insurgents was in charge of some area. Al-Qaeda had their own patrol groups driving around making scary noises to terrorize neighborhoods and let them know they were there.

Fear made me very cautious. Sometimes if a door was opened, an entire family was killed—old people, women, and children. I had four huge locks on the front door. It looked like a protected holy place. I would only open the door for family. We had a way of knocking that was a signal.

Al-Qaeda started targeting college graduates. My father-in-law is a former [deleted] and a college graduate. We are an educated family. I had a family member at the government office. He offered me a job. So I covered my face and went for the interview. Women were not allowed to drive, so I taught my 11-year-old son to drive. He took me to work. He had a lot of accidents. He also used to take me shopping.

Three or four times I was stopped at al-Qaeda checkpoints. They asked, "Where is your husband? Tell him to come in or we will kill him." I started crying and told them that he ran away after he lost his job and that we are separated. So they let me go. The al-Qaeda commander in our area started watching me himself from a vehicle outside the fence.

I used to take my younger son walking. Females were not allowed to walk by themselves. They had to have a husband or male relative with them. So my son had to go with me.

Al-Qaeda controlled all fuels. The al-Qaeda group commander refused to allow them to sell fuel to me. I covered my face, but he recognized my eyes.

I felt like I had no life in Ramadi. I used to bring hairdressers home sometimes. But al-Qaeda asked why a male was in my home if my husband was not there, so I had to stop.

They were not mujahedeen; they were low-life people. A true enemy will face you and fight. These were cowards, cheaters, and backstabbers. They threatened my son at school. I told him, "If you see their cars, jump the back fence of the school."

Many of our original neighbors fled the city, and the insurgents lived in their homes in [other] areas. We are thankful for [name of prominent person, deleted]. He deserves a lot of credit. A friend of my in-laws heard of our situation and told [him]. He sent a force to get us out. He hired my husband back as a police officer. He worked six months at a station for [title deleted]. My husband lived in the station while I lived at home with the kids.

I was afraid of the Coalition forces and the insurgents. I used to leave the house at 6 a.m.. Before 7 a.m., there was a Coalition forces curfew. After 7 a.m., the insurgents were out. I used to carry a white flag with me. It was my husband's undergarment. The Coalition forces know the white flag. Nothing works with the insurgents.

This is the first time I've told my story without crying. I am happy. I just want to protect my family and love them. My husband and I were in love before marriage. That is rare here. Most marriages are arranged. If they killed us, I wanted them to kill me first. I can't bear to see him dead.

Our neighbors were on watch and asked questions. The neighbors were collaborating with al-Qaeda. Sometimes I disguised my husband as a woman or an old man. Every day was a different story. It was like living in a different movie every day.

The neighbors were asking questions, so we moved to a different neighborhood. I hid my husband in the back of my car. I took out the seat and put a blanket over him. There was no room to park the car at the new house, but our neighbors were so kind and let me park the car in their garage. The husband turned out to be a big insurgent. I told them that my husband was gone. I never told them the true story.

I asked my husband's boss to move my husband to [a different place]. Life was better there. Things were running smoothly there, and Ramadi was a ghost town.

The Awakening started with Sheikh Sattar Abu Risha because al-Qaeda killed many of his family. The police started coming down slowly. People started coming out.

We had the first woman's meeting on December 15, 2006, at the [place deleted] elementary school. Sheikh [name deleted] attended the meeting. He paid attention to the women's movement.

Captain Tina [pseudonym] of the Coalition forces was present at the first meeting, and she worked with us. We set up literacy classes for females. I worked with her, but I was also afraid. I stayed in the house for a week after the first meeting. I had a photo taken with Captain Tina, and it was posted all over town. She e-mailed it to my organization, and one of the women in our office let it leak out. Now women are going back to have pictures made with the Coalition forces.

After Captain Tina we got Captain Stephanie, and we met at the elementary school. Some women refused to shake hands with Stephanie, but she is very patient. She tamed them like taming wild animals: she brought them gifts and listened to problems. She won them over slowly. Stephanie opened literacy programs and classrooms at three places.

Stephanie is a great American woman. She taught us how to depend on ourselves. She gave us a financial mind. Our mindset was to get a degree, get a job, get married, cooking and kids. Stephanie taught us to do finances, fight gossip, depend on ourselves, and fight the insurgency. For example, a woman can't read. Her son comes in with money and says that he is on a holy mission. We should ask questions like, "What kind of holy mission? Who is telling you this?" The first thing is to teach kids morals.

Women who didn't like Stephanie are now trying to pull her away to their side. Stephanie distributes products. We call her "Santa" or "Mamma Claus."

Stephanie helped people love security. She helped women get jobs. She put rules on who should be hired: target unemployed college graduates to maximize employment.

In 2007, we organized as a united women's organization. We met at Sheikh [name deleted]'s house and discussed having more literacy classes. It's the only way to open the female mind. And women need to get more involved in their children's lives.

We opened our first classroom at a school. Stephanie asked us to make it succeed so the public would accept it and it would expand. We gathered new members in Ramadi, but the women were scared. We started with friends, and some came from Fallujah. We succeeded and now have 400 members in different parts of Anbar.

At the time, it was raging with insurgency. There were no rations available, except through Stephanie. She brought in a truckload of food and supplies—1,500 shares. Stephanie asked me to hand out the clothes today so that the kids could wear them tomorrow. I was fasting and it was hot, but I spent 12 hours handing out clothes. Some of the kids were so excited that they changed clothes immediately.

Stephanie is really great, comes all the way from America. I don't fear anymore. Stephanie showed me the way, how to love the responsible role that I take. She taught me not to fear but still be cautious.

We help women get jobs—jobs for widows, cleaning courtyards. If a woman can't make it, her kids do it, so that she still gets a paycheck.

The first meeting was with Donna [surname deleted] in [place deleted]. She asked for the number of widows. I gave her a record of each widow and the number of kids. We met with the director of social services in Ramadi. We identified the greatest need and formed a project. She was replaced by Lieutenant Carla [surname deleted].

Stephanie built women into leaders and gave them confidence. She inspired everyone. We are getting more organized and getting better government support. But we still need more support for the orphans.* There are 850 orphans just in Ramadi. Families shelter them, but they need help. We're collecting backpacks, mittens, shoes, caps, and sun block. We need classes in nursing, English, cooking, computers, making artificial flowers, painting on ceramic. We need a new hospital. There is no room in the existing hospital to deliver babies.

Iraq once had female doctors, judges, and scientists, and they were also good wives and mothers. We want you to separate us from Saudi women and the Gulf States. Twenty-five years ago, we had an Iraqi female minister. The world thinks all the women are like the ones in Saudi Arabia and other countries. We are not like that. Before Saddam [Hussein], women were free to get out. We had equal education. We're coming back. The wars, the insurgency, the Sunni and Shi'a conflict were a setback. The people who started the Sunni-Shi'a conflict were not Iraqis.

Chief Warrant Officer-4 Timothy S. McWilliams: Are you in touch with the rest of the world?

Miriam: We have satellite TV now. We're careful about what is on. Not all of the programs are good. We have the Internet. It's expensive though, about $50 a month. We're interacting with the rest of the world now. It's a sign of civilization. Iraqis are conservative but not extreme.

---

* In al-Anbar, the term "orphan" includes children with only one parent.

Religious Perspectives

Interview 2

Dr. Thamer Ibrahim Tahir al-Assafi

Muslim Ulema Council for al-Anbar
(Council of Muslim Scholars)
Professor of Religious Studies, al-Anbar University
Senior Theologian to the Sunni Endowment
Ramadi City Council Member

Sheikh Abdullah Jallal Mukhif al-Faraji

Head of the Sunni Endowment for al-Anbar
Regent Sheikh of the Abu Faraj Tribe
Ramadi City Council Member

Dr. Thamer Ibrahim Tahir al-Assafi is a cleric in Ramadi who chose a religious path after serving as a commando in the Iran-Iraq War. He is influential in Iraq and well-regarded among Muslim endowments in the Middle East and Southeast Asia. His nexus of influence is al-Anbar University, where his credibility is magnified by his combined status as a member of faculty, combat veteran, and cleric. He was an early proponent of engagement, and throughout the insurgency he used his sermons in the mosque and his lectures at the university to publicly call for moderation and restraint. He was shot in the shoulder during an assassination attempt against Governor Mamoun Sami Rashid al-Alwani, but he continued undeterred.

Sheikh Abdullah Jallal Mukhif al-Faraji is a former Olympic athlete, a cleric, and one of the original members of the Anbar Salvation Council. He is addressed as "sheikh" as an honorific. He is neither a true lineal sheikh nor a pretender. The true sheikh delegated most of his responsibilities to Abdullah, who is popular among his fellow Faraji tribesmen. His influence is concentrated around Ramadi and eastward toward Fallujah. During the insurgency, he used his influence to temper the insurgent movement, anti-U.S. rhetoric, and infighting among the various militias. He was wounded several times, including a gunshot to the arm during an assassination attempt against Governor Mamoun. The governor seems to regard him as a personal spiritual advisor and trusted agent. While recovering from the gunshot wound to his arm, Sheikh Abdullah was mistakenly arrested by Coalition forces. Brigadier General James L. Williams was instrumental in securing his quick release, an act which established a greater degree of trust and resulted in more Sunni leaders entering the political process.

Dr. Thamer and Sheikh Abdullah were interviewed by Colonel Gary W. Montgomery and Chief Warrant Officer-4 Timothy S. McWilliams on 12 February 2009 at Camp Ramadi.

Dr. Thamer Ibrahim Tahir al-Assafi: History speaks for people, for communities, so from us it necessarily is required to write down history. Whether we like it or not, we have been joined with you in history. We have to clarify for distant people and for later generations.

There is one overriding fact that you will have to know: there was no terrorism in Iraq prior to your coming here, except state terrorism. After the invasion of Kuwait [1990-91] and the total embargo that was placed on Iraq, it created a generation of people that was destitute and corrupt. They would be taken to court, and they would be imprisoned, and most of them were angry. Just before the time of Baghdad's fall to the Americans, Saddam [Hussein] freed all these prisoners in a general amnesty. Those people were set free among society.

After the fall of Baghdad, you noticed there was chaos, and you noticed that government institutions were looted, and particularly after the dissolving of the Iraqi army by [Ambassador L. Paul] Bremer. The internal security apparatus was dissolved, also. So there was a general freedom, and there was freedom of movement.

Saddam also opened the borders to allow foreigners to come in before the battle with the Americans, and a lot of them did come in. Most of these people were Syrians. Most of these people gathered in our province. We tried to get rid of them. We gathered a lot of money, and we hired vehicles, and we tried to kick them out. But they went out to the surrounding villages and hamlets, and there they organized themselves into cells under the guise of resistance to the Americans and the occupation.

After the fall of Baghdad, which was on a Wednesday, eventually we established a government in Ramadi with a new police chief. He chose new policemen. These policemen, some of them were bad elements, from among the people that Saddam let out of prison. I raised the issue with the police chief, and I asked him, "Why would you put bad elements in charge of the community?" He said, "Look, when I bring these people in here, I'm in charge of them. I keep watch on them, rather than turn them loose in society to create mayhem."

Negotiations ensued between the tribes of Anbar and the American forces. Coalition Forces were located primarily in Rutbah and the west of Anbar, and they wanted to come in from Baghdad, and they wanted to come in peacefully. An agreement was struck between some of the tribal sheikhs and the American forces for a peaceful entry. After they entered Ramadi, there was a big demonstration, a peaceful demonstration, because they did not approve of an occupier coming into their capital. The American forces did not respect the people who were demonstrating. They dealt with them rather violently. The people's reaction was to pelt the Americans with rocks and tomatoes, and it was a rather negative reaction. They provoked the citizens. That was the first thing that started hatred.

The next day, they demonstrated again, and the Americans treated them in the same manner, meaning their armored vehicles went right through them. A young man, an 18-year-old youth, threw a rock at an American tank, and the soldiers shot him dead. We are a tribal people, and in our tradition, we know revenge. If someone gets killed from your family, you have to kill the killer, or at least a relative of his.

When the Iraqi army was dissolved, they left a lot of armaments, including armored personnel carriers, heavy machine guns, and a lot ordnance. People took them and hid them in their houses, not to have a future confrontation with the Americans, but in fear of a confrontation with Iran. Keep in mind we are military-trained people as a society because of the battles—the Gulf War, the Iran War, the Kurds in the north. Most of us were in the army, so using weapons was something we could do with ease.

So these people whose youth was killed by the Americans, they formed a cell, and they started looking for revenge. They found out that placing an IED [improvised explosive device] is a simple matter, so a lot of cells began forming all over the place.

When the foreign Arabs came in, they came in with suitcases full of dollars, and they started organizing cells. They got in touch with the Iraqi people, and they started organizing them better. We had a weak police force, a weak army.

The Americans did not want to get involved unless they were directly attacked. If an American patrol was on the highway, and they saw a dead person, they would just leave him there. And this really started to create hatred toward the American GIs, because they couldn't care less what happened to Iraqis. If they were killed right in front of them, they did not get involved. So that reaction of the Iraqi people was like, we hated the Americans.

When the terrorists attacked the national mosque in 2005, Sheikh Abdullah was attacked, I was attacked, Sheikh Tarek was killed—the head of the Sunni Endowment. The American forces were 200 yards away. They were watching and did not interfere.

There was a gap between the American forces and the Iraqi people, and there was strong hatred. They did not improve the police. The police essentially got dissolved. There was only one unit, the highway patrol, and they stayed to themselves. They stayed in their headquarters. . . .

We had a huge conference with the American commanders and the tribal sheikhs to develop the Iraqi police and Iraqi army. In 2005, we had people gather to volunteer for the police, and two terrorists infiltrated the place and blew themselves up. The American forces did not even give first aid to the wounded. And the civilians who aided those wounded, the terrorists would catch them down the road and slaughter them. Another conference was held, and we interceded with them to develop the police and the army, and, again, they intervened. The terrorists said we deserved death, and many among us were killed.

Being clerics, we used to incite people against the terrorists through our sermons in mosques and speeches. The last election, the one in 2005, we defied the terrorists, and we went out to cast our votes. Afterward we were subjected to the wrath of the terrorists for the people that went out and voted.

Had the American forces allowed us to carry arms at that time, we would have eradicated terrorism much sooner. Any of us who carried a personal sidearm was arrested, while the terrorists had plenty of heavy armaments and were walking freely. So we were lost, and the American forces did not stand with us. There were many people that got killed by the terrorists, and the Americans did not stand with us, so our backs were open, so to speak. We could not get help, not even from the Iraqi government. Life became very difficult, because the banks were looted by the terrorists.

In the meeting, I accused the commander, [Brigadier General James L.] Williams, of being in cahoots with the terrorists, for looting the banks. I said, "Look, how can the terrorists go into a bank and take the money out, and you're right there—and you're not in cahoots with them?"

The province of Anbar became so paralyzed that even the employees of the banks could not receive their own salaries. Many fatwas were issued by the terrorists to close the universities. Ramadi became a ghost town. Universities, schools, factories, and institutions were all shut down, so we were pressed. The government did not pay us our salaries. Life became intolerable. So we started looking for salvation, no matter who it was.

Salvation came through the Anbar Awakening with the leadership of Abdul Sattar Abu Risha. I tell you frankly, we all disliked Sheikh Abdul Sattar Abu Risha. This being because he was not a religious person, so there was a distance between us. When Sheikh Abu Risha proposed the Awakening, we looked into it deeper, and we found it to be a really good thing to do. So I got Sheikh [Abdullah] Jallal [Mukhif al-Faraji], and we said this is the thing to do. We have to do it. And so we carried on, and with the help of Colonel [John W.] Charlton [USA] and Lieutenant Colonel Warner, we were able to get rid of the terrorists. The religious establishment started working hand in hand with the Awakening. When they liberated a street or a town, we used to go in and open the mosques that were closed by the terrorists. Some of the mosques were also closed by the American forces.

We lost many innocent people. Some of them were loaded down with blocks and put into the river. Some of them were taken out to the desert and beheaded. Some of them were imprisoned.

We were hoping after the Awakening—after getting rid of the terrorists—we were hoping that new building would take place, but it hasn't happened. A lot of money was paid out by the American forces to many contractors, but nothing came of it; nothing useful, that is. The streets that were paved got destroyed. They did a lot of painting of buildings and shops. We were not exactly happy about what these people did, all this money that was stolen. All the money was given to affluent people—tribal sheikhs and contractors—which could be considered as a sort of bribe.

We were hoping for the civilization of America to be transferred into Iraq. Maybe you're not a colonial power; this is new to you. We're hoping for many institutions. When Great Britain occupied Iraq, they left many institutions, bridges, hospitals. People can go right now and say, "There is the bridge that was built by the British. This is a hospital built by the British." When you leave Ramadi, or Anbar all together, what will your legacy be? It's total destruction. People will say you just came in, destroyed, and left. You're going to help us rebuild our country? A translator asked me a while ago, "If I come to Ramadi on vacation or to spend some time, where do I go?" I told him, "The street, where else? There are no hotels, no motels, there aren't even parks to go into." This is all I have to say.

Colonel Gary W. Montgomery: What was your impression of the foreign Arabs?

Thamer: We totally reject the foreign Arabs because we have prior knowledge of their doings in Afghanistan, in Chechnya, in Bosnia-Herzegovina, and in Algeria. They were hired hands, and Iran hired them. We knew that beforehand, but the security situation was really bad when they came in. The Ba'athists harbored them. I was just joking.

Sheikh Abdullah Jallal Mukhif al-Faraji: The most glaring mistake committed by Coalition forces was leaving the borders unattended. From these borders, the foreign Arabs entered. The second most glaring mistake they made was dissolving the army. Initially when the Americans came in, there was a peaceful coexistence between the people and the Americans. There was a great understanding between the tribal sheikhs in Anbar, especially [Ali Amar Limand], and there was no tension.

The foreign Arabs, when they came into Iraq, they changed the mindset of the Iraqi people and told them to use bombs.

I see things from a different perspective than Dr. Thamer, so I have a disagreement with how he sees things. I don't want Dr. Thamer to forget al-Ghazaliyah, Baghdad. He and I and our peers joined with the Americans. We killed four terrorists in Baghdad, so I don't want him to forget that they helped us in 2007.

Thamer: Yes, that's true, but that's after the inception of the Awakening. If it were before the Awakening, they would have killed us. . . .

Abdullah: And I liked Abdul Sattar Abu Risha, even though Sheikh Thamer hated him.

Thamer: Look, I hated him before the Awakening. We even met with him, and I told him personally that I hated him before the Awakening. But since the inception of the Awakening, no, I liked him.

Abdullah: Truthfully, Iraq had no foreign Arabs before. The reason the foreign Arabs came into Iraq is because they were following their arch enemy, the Americans. It's apparent that the United States will fight these foreign Arabs all over the world, and when they came into Iraq, that's where the battle took place, right here in Iraq.

Now, the American forces assaulted some Iraqis in their houses, and everywhere else, and that created a negative reaction toward the Americans. To be truthful, the U.S. forces didn't distinguish between who is a foreign Arab and who is an Iraqi, so they shot everybody. . . .

I represent the Sunni Endowment Organization before the Awakening and before the religious people. This is, right here, the telltales of them shooting me. Right here. He [Dr. Thamer] got shot, too. The governor, Mamoun, got shot. We were in the same building when they attacked. We took their own weapons, my force, and we fought them with their own weapons. I don't want to talk too much, but now I want to say this is the inception of the Awakening.

What Dr. Thamer explained is sufficient enough. We came in as the religious side with Sattar. We gathered this bridged duality of the religious people and the tribal people, and they were bound together. When finally the people realized that the clerics were in with the Awakening, they came in and joined Sattar. That's why the Awakening succeeded. . . .

And one cannot forget [Lieutenant] Colonel [Thaddeus] McWhorter [USA], [Lieutenant Colonel Michael] Silverman [USA], and Charlton. . . . They knew how to treat the tribal people. They knew how to treat educated people in the cities, and village people. They trained their minds, their military minds, their civil minds, to understand all this. Please, if you write down history, do not forget these people. I'm a religious person who speaks only the truth, and I don't know who is who, but I just tell you the facts as they are. They've established a great legacy for the Coalition forces. They were loyal to their country; they were loyal to their military.

Abdullah: I want to talk about politics, reconstruction, education, and economics. This election taught Iraqis to understand what an election is. There is a political awareness among the Iraqi people.

As far as education goes, we need an exchange of ideas, of views between us and the other religions, and not just religions, but civilizations. We've demanded the establishment of a university, an American university in Anbar, or anywhere in Iraq, so there will be an educational exchange between the United States and Iraq. We've demanded the establishment of libraries that will have all kinds of books—American, Arabic. We have demanded the establishment of an English-teaching institution, because we need these institutions. There are promises to that effect. We need an educational institution to rid the people of all the negative ideas in Iraq and the surrounding areas. We need a monument left for the American forces in Iraq, a hotel, a bridge [conversation with interpreter].

Interpreter: Again, he is speaking about the bridges that the British built.

Abdullah: When you cross over it, you say, "This is the British legacy." We need something to establish your legacy. What have you left? Sometimes it's paint.

As for economics, a lot of money was wasted, as Sheikh Thamer said, to specific people. About four to seven days ago, U.S. Forces Command in Iraq spoke of these financial mistakes. And the chief cause of it is corruption, whether among Iraqi society or the U.S. forces themselves.

At the end, I will wish you the best in writing history, but we don't want to forget what you went through.

Colonel Charlton knew how to treat people. When a person came to him raging mad, he used to smile in his face, calm him down, and deal with him positively.

Colonel Silverman knew how to treat Arabs, because he lived in the Arabian Gulf. So when they put him here in Iraq, he knew how to treat the Iraqi people, and he treated them well.

One other person is Major [Adam T.] Strickland, and another guy who called himself [name deleted for security reasons]. He was an American. Those people knew how to treat Iraqis.

Not to forget General [John F.] Kelly's role, because he used to shuttle between the tribal areas. We can never forget it. He knew how to treat people and to win them over. He would release detainees. He honored sheikhs. He would release a detainee as an honor to sheikhs. He supported those who are starting service projects.

They knew how to connect society with the rule of law. Everything about the military, they knew how to connect it. To be honest with you right now, I cannot think of everything good that you've done, because it doesn't come to mind.

And we cannot forget Brigadier General Williams, who I have mentioned, who was here before the Awakening in 2006. . . .

You must know the period between 2003 and 2009 is not an easy period. It was a very complex period. I want you to know that some of the clerics did side with the terrorists because they had religious tendencies. So we talked to them, and we told them the right way, and we won them over.

Sheikh Sattar even sent me to Oman to talk to Sheikh Harith al-Dhari and to tell him to stop all the fiery speeches, the negative things about the police and the army. We calmed him down, and we won him over, and no more fiery sermons. We said that we want security—the army, the police, everything—in our own country. We said, "Sooner or later, American forces will leave our country, and we will have to stay and rebuild it." He was somewhat satisfied with our suggestions, so for a year or two he did not say anything negative.

Montgomery: What part did al-Qaeda in Iraq play in the troubles here?

Abdullah: Their role was simple. They are the primary mischief makers in Iraq. They did not know what resistance meant. They did not know what jihad meant. They did not know what humanity meant. And I speak in front of the camera, and I say it forcefully. They did not know what life meant. They wanted to kill everybody that existed and everybody that did not agree with them. So, therefore, [inaudible], and that's why they had to be kicked out of this country and any other country. And I'm not saying this to be nice to you. And, if you get me madder, I will say much more. So, please, leave me be.

Montgomery: I think that we definitely don't want to get him angry.

Abdullah: I'll tell you what they used to do. They used to kill the doctor, who is in the business of bringing people back to life. They used to kill the children. They used to kill the women. They used to kill the engineers, whose job is to rebuild. How can you kill everybody?

You ask them the question, "Why do you kill?"

They said, "Well, this guy's not a Muslim."

"Okay, why don't you call him Muslim?"

"Oh, because he's a Shi'a, or because he's Kurdish, or because he's a Sunni backslider."

They had a reason for killing everybody. "This is a Sunni backslider. This is a Shiite. This is a Kurd. This is a policeman. This is a soldier." So they had a reason to kill anybody and everybody they wanted to. We tried to understand who the real enemy was. We just didn't know who it was. They consider a little baby, a woman, to be an enemy?

You may not know this; perhaps you do. The religion of Islam prohibits the killing of a child or a woman, even the cutting of a tree or destroying a building. This is not done in Islam. Islam means peace—peace and mercy for all people, not just for Muslim people. They used Islam as a cover, a pretense. I, being the head of the Sunni Endowment organization in Anbar, I say, and I declare it to everyone everywhere. And I would like to tell you, there are very few, and soon enough we will be rid of them. It is a dark spot in a bright room.

And, you know what? We can say frankly that there are some Christians who destroy the Christian religion also. There are very few of them, but they do exist. It's not only in Islam, but even in Christianity, though not the slaughter. Many sects pretend to be religious, but their aim is to kill people.

I said what I said, and I ask God for forgiveness. I wish success for everyone. I thank you for this meeting. And I'm ready and able and willing to start writing history as it is. History will record the bad deeds of the people who did evil. And good words will be written for those who are good people.

Tribal Perspectives

Interview 3

Sheikh Ahmad Bezia Fteikhan al-Rishawi

Paramount Sheikh, Albu Risha Tribe
President of Mutammar Sahwat al-Iraq
(Iraqi Awakening Party)

Sheikh Ahmad Bezia Fteikhan al-Rishawi is a brother of the late Sheikh Abdul Sattar Abu Risha, who was the founder and iconic leader of the Anbar Awakening Council. After his brother's assassination, he was elected to succeed him as paramount sheikh and president of the council. He was interviewed in his home in Ramadi, adjacent to the guesthouse where the Awakening was announced, by Colonel Gary W. Montgomery and Chief Warrant Officer-4 Timothy S. McWilliams.

Sheikh Ahmad Bezia Fteikhan al-Rishawi: We took the Iraqi commander to the American forces at K-160 [kilometer marker 160 on the highway], and we told them he was ready to talk. They said, "Bring him forth." Sheikh Khamis* wanted a guarantee from the American forces that if he brought the Iraqi commander, he would not be apprehended. They said, "No," and that was a promise.

Sheikh Mohammed went along with them to K-160. Brigadier General Ahmed Sadag also went. They signed a peace treaty—no need for fighting—and he promised them he would discharge his soldiers. So the American forces entered Anbar with no fighting. That was 2003. A government was established in Anbar with the support of the American forces. Basic services, and everything else, were excellent.

Some people started inciting people to fight the Americans. They described those among us who cooperated with the Americans as stooges, and they gained a lot of popularity. They marshaled the people against us, so we left them. Right then and there, they started attacking the government of Iraq, the Iraqi police, and the American forces. And then, at that time, al-Qaeda came in. And that resistance, the foreign Arabs, the emirs, came in place.

---

* Possibly Sheikh Khamis Hasnawi al-Issawi, paramount sheikh of the Albu Issa tribe.

They killed a lot of personnel from the police. They destroyed the Iraqi army. They destroyed the infrastructure of the country. They attacked schools. They attacked university professors. They forbade involvement in any political dialogue. The situation became unbearable. The sheikhs and the brains left the province—professors, teachers, and doctors—they all went to Jordan and Syria. Al-Qaeda roved the country. We set up security forces around our compounds, and we stayed here. We did not migrate. They started lobbing mortar rounds on us in order to make us run away. We did not. We kept being friends with the Americans.

Infrastructure services were really bad. Education and health care were bad. Al-Qaeda took over the Ramadi hospital, the Department of Education, the university, the oil infrastructure, and they were the rulers. The governor of Anbar was protected by the American forces. Not many people stayed with him, except a few accountants who were working on the budget. Some of the budget was going to finance al-Qaeda via some contracts or projects. It wasn't given directly to al-Qaeda; it was channeled to them in various ways.

We realized that the people had had it with the situation. People started murmuring, saying that if we had obeyed the people who started work in 2003, the situation would not have deteriorated to this point. So they started the blame game, due to the losses that were inflicted among their ranks and their tribesmen. So Sheikh [Abdul] Sattar [Abu Risha] and I, we started thinking that we've got to get in touch with the tribal sheikhs and their cousins, the ones who were active, so we could incite them to fight al-Qaeda.

Sheikh Sattar told me, "Leave it to me. I'll take care of it." So he started moving, talking with the tribal sheikhs, one by one. He told them that he was ready to do something, and he gathered them for a conference on the 14th of September 2006. A communiqué was issued, containing 11 points:

First, to bring back the army in Anbar and to bring the tribal sons into the police and army.

Second, to declare war on al-Qaeda—and he described them as thugs and criminals.

To bring back the respect that is due to the tribal sheikhs.

The American forces were to be considered friendly, and attacking them was forbidden. That was the first conference ever held in Iraq with the American forces to be considered friendly.

To treat the Ba'athists humanely.

And no cooperation or negotiation with al-Qaeda.

To reopen the judiciary, and to bring the criminals before the law.

To be presented as government officials, as people who follow the law, and not as militias.

To enter the political system, and to enter into dialogue on a large scale, and to participate in the election.

These were the most prominent points that were announced at the conference.

Also, getting in touch with the Iraqi central government will be considered as reconciliation. And the formation of a council, which was called the Anbar Salvation Council, to take the place of the provincial pouncil that ran away to Syria. The central government refused that request, but they put eight people from the Salvation Council into the provincial council.

The central government agreed to let the people of Anbar enroll in the army and the police. The war on al-Qaeda started right then, and we began to take their men before the law. The American forces started organizing the police by taking their applications, screening them, training them, and, whenever we had a ready force, we established a police station with them. American forces helped us with training, with salaries, with transporting salaries, and with arming the police.

The war against al-Qaeda was waged first in Ramadi, and then in every other tribal area. Whenever we cleared a tribal area of al-Qaeda, we set up a police station. We elected a representative company from every area. We took these people, and we went to the city of Ramadi, and we attacked in Ramadi. That's how we liberated Ramadi. Tamim was also liberated. Malab was liberated. Albu Fayed was liberated. Tajiria, Khalidiyah, Amariyah, Fallujah, Saqlawiyah, Karma, Jazeera, the outskirts of Hit, Albu Issa. The city of Hit proper was liberated. Baghdadi city was liberated. The city of Haditha and all the adjacent territories, Rawah territory, Ana. Al-Qaim was liberated by the Albu Mahal tribe. They started before the inception of the Awakening. Al-Qaeda started a fight with them, so they counterattacked with their sons, and they were able to liberate their area.

We opened a new page with the tribes in the south. After the fight with al-Qaeda and the victory against them, we renounced sectarian violence, so we invited tribal sheikhs from the south to Anbar. There was a dispute among the Shiites and Sunnis, and we wanted to alleviate that. We wanted to cooperate with them as one people, as Muslims.

We opened up the highway. We put many checkpoints on the highway. Those people who were intercepting travelers—the criminals—we started fighting them, and we detained many of them and put them before the law. As you see, the highway right now is the most secure road in all of Iraq.

Tharthar, which contained many mass graves—the Tharthar area was attacked by the emergency response units. It was liberated. The Samarra road.

Al-Qaeda was broken, with the help of God, and they will not be able to conduct any military operations. You hear, now and then, about VBIEDs [vehicle-borne improvised explosive devices] or someone committing suicide. This is the choice that we gave them: you can either surrender to us or you can be suicided. They despaired of accomplishing any victories, so now they resort to suicide.

The organization concentrated on my brother, peace be upon him. They conducted 12 suicide attacks against him. The twelfth operation finally hit him exactly a year after the inception conference.

We changed the Awakening into a political entity. We became a political movement. We brought in people with high degrees: tribal sheikhs and their sons and relatives, tribal sons from the educated people who are able to lead, and we concentrated on them being clean and capable. We waged a campaign for the election, and we achieved significant progress. We encouraged the people to go and cast their votes. We encouraged the people to go and participate. We emphasized to them that their vote was very important and crucial at that juncture. We encouraged them for change. We obtained very good results in the election. We surpassed all the parties that used to be dominant. We overcame them. After the election results are announced, we will form alliances with the people that garnered enough votes to have a majority and form a government. This is the Awakening. . . .

I used to pay salaries to the police, who did not receive a salary for four months. Uniforms were imported from Jordan. I did not want them to look like militias by wearing civilian clothes. I wanted them to look official. We were careful not to conduct any military operations in civilian clothes, only in military uniforms. The martyrs who fell in battle, we used to go to their funerals and help their families monetarily. Weapons were short. We didn't have enough weapons, so we had to purchase what we could. These expenditures came out of our own pockets. At the time, we did not receive any support from the government of Iraq or the American forces.

Colonel Gary W. Montgomery: Where was the conference where they came up with the 11 points? What was the location?

Ahmad: At this house, right here.

Montgomery: So in this building we were just in. And I understand all of the tribes freed their own areas, their own tribal areas, but it happened in succession. Once a tribal area was free, did that tribe go to help another tribe, and so forth, or did they just each separately clear their own areas?

Ahmad: No, they elicited the help of other clansmen, from other tribes. But once they liberated a place, they left people from that tribe to be policemen in that police station.

Montgomery: So they did both. They protected their own area and also . . .

Ahmad: And helped other tribes. When we went out there, and we inquired about who was in al-Qaeda, we got a lot of help from the tribe's people in the area.

Montgomery: When they met with the Shi'a to get rid of their differences, who did they meet with? What was that like?

Ahmad: With many other sheikhs, al-Zargan sheikhs, al-Fetna sheikhs. They wanted to keep Iraq united, and they wanted to avert any internal conflict. It was a successful meeting.

Montgomery: Was that here, also?

Ahmad: At the guesthouse. No one was willing or able to undertake such an endeavor, to fight al-Qaeda in Anbar. No one was even willing to stand before the television camera and announce that he would fight al-Qaeda. We announced, right from this house, an open war against al-Qaeda, right on TV.

The organization got wind of it. They were right there. They had many weapons. They lobbed a lot of mortar rounds on us to get us to back off, but we never did. We never wavered.

Unknown: What actions did he think could have been done differently in the very beginning to possibly start this sooner? Was there anything either the Coalition forces, or the chiefs, or even the local government could have done to possibly get this going sooner?

Ahmad: As the American forces, per se, you could not have done anything. I'll give an example. If a contractor took money from you to build a school, he was killed because he was a collaborator. He says, "Look, please, look, I built a school." They said, "Yes, but you took money from the Americans," so they killed him. Therefore you all would have been very limited, if not at all.

As for us, we were able to convince the people that the situation was dire, and we needed to rise up against al-Qaeda and live like normal people, like advanced people. When we used to meet with the tribal sheikhs, we asked them—we lectured them, "Look, when are you going to have a normal life? When are you going to be able to get your dues from the government? When are you going to be able to send your children to school? When are you going to live a normal life, when these thugs are in charge? These people are wearing masks. They cannot build the country. People that can build government and sustain it are the police and the army. You cannot get your rights if you have no police and no army to protect you."

Unknown: You said that the Coalition forces had a hard time distinguishing, sometimes, between the Iraqis and al-Qaeda. How easy is it for you to distinguish? Are they still hiding out here, or have they left?

Ahmad: It is very easy for us to distinguish, and there are no foreign Arabs here now. It is hard for you to even distinguish between the guilty and the innocent, but we know them by their faces. This is a criminal, this is not. Some of the mistakes you made were in Bucca prison. It was better to have referred those criminals to the Iraqi justice system than to Bucca, because they started having a school, and they indoctrinated other detainees.

Interview 4

Sheikh Wissam Abd al-Ibrahim al-Hardan al-Aethawi

Former First Deputy
Sahawa al-Anbar/Sahawa al-Iraq
Current President of Iraqi Popular Front

Sheikh Wissam Abd al-Ibrahim al-Hardan al-Aethawi is a member of the al-Hardan clan, which is the paramount family of the Aethawi tribe and among the most respected families in al-Anbar and Salah ad-Din provinces. He was a member of the Tribal Affairs Council and among the founding members of the Anbar Salvation Council. He is not influential by way of a large following; rather, because he is from a very prominent family, he knows many important people and knows the background of many important people and significant events.

Sheikh Wissam was interviewed by Colonel Gary W. Montgomery and Chief Warrant Officer-4 Timothy S. McWilliams on 15 February 2009 at Camp Ramadi.

Sheikh Wissam Abd al-Ibrahim al-Hardan al-Aethawi: After too many killings had taken place, to the Iraqis . . . the nationalists rose up. I was the head of that movement. We gathered about 500 army officers from the last army, and we rose in 2005. I went to Sa'doun al-Dulaimi. He was the defense minister at the time. He was my relative. I asked him to give the fighters permission to carry arms. I was not successful in that endeavor. We could not get government's permission for them to carry arms.

Most of the officers left the country because they were subject to being eliminated. They were traced, and they were killed by the terrorists, many of them were. One of their leaders, Muhammad Ishmael al-Jabawi, was eliminated. He was from the previous Iraqi army. He was a leader in the national resistance. A general in the air force, Waji, was also killed. My brother, Mohir al-Hardan, who was a judge, was also killed. He was a deputy to the governor. Sheikh Naser [Abdul Karim al-Miklif] was also killed. Sheikh Jai

Tehan Abu Risha was also killed, and many more notables—nationalists who would serve the country—were eliminated.

Colonel Gary W. Montgomery: These were all outside the country?

Wissam: No, inside the country.

Montgomery: Inside the country?

Wissam: So those who remained immediately left the country.

Montgomery: Right, okay.

Wissam: The militias also had a role in making people flee the country, based on sectarian divide. Some people were displaced. So the last of the national resistance movement that I envisioned was left with very few people. You could count them on your hands.

I don't want to make the story too long, because if I am to talk about the Awakening and its history, it would take months. Sheikh [Abdul] Sattar [Abu Risha] called me by phone. He said, "Please, I want you to come to Anbar." I was living in a village. The name of the village was al-Mish Haniyah. That's where my grandfather lived. It's about 70 kilometers from here. I asked what he had in mind. He said there was a conference about to take place in Amman, and I told him I did not wish to attend any conferences, because they are to no avail. He insisted that I attend, or more likely, he pleaded with me. His cousin is my wife, so he kind of pressured me into attending. Sa'ad Hamid al-Douish, his cousin, also was martyred. His mother is also from the same family, the Tehan family. I brought Ali with me to Anbar.

When he saw me, when he laid his eyes on me at his house, he said, "We are victorious by the God of al-Kaaba." When I asked him what he had in mind, he said, "We have a huge undertaking, and nobody can take care of it, except you." Before he even said hello, which is customary in the Arab world, he started telling me what he had in mind. He said, "I want to fight al-Qaeda."

I said, "How many do you have?"

He said, "I have 7,000."

I knew for a fact he had maybe 70, not 7,000, but I told him, "Your 70 would be considered as 7,000."

So all night long, we sat down, and we were planning and plotting for this huge revolution we're about to undertake. My plan was as follows. The first phase of the plan was to start an educational program toward the tribes. The second one, when the people finally saw that I was involved in it, they became courageous.

It was hard for the tribal people to come into Abu Risha's compound because he was considered an agent of the Americans. When I sent for groups to come in, a lot of times, out of four you may get two killed on the road. That's because too many terrorists were about on the road looking for people that would cooperate with us, and they would eliminate them. The difficult places, both topography-wise and infested with terrorists, were Albu Aetha, Albu Faraj, Albu Beit, and Albu Bali.

I planned to make a nucleus of 50 men in each tribe. Those 50 men would be ready after we went to the conference in Jordan and announced the revolution against the terrorists. They would be ready to move right then.

Okay, this meeting started with me and him in the evening. It ended up in the morning with 10 more people. The notables would be Hamid al-Hayess, Shalan Nouri, Jabar Ajaj, Muriad Maieshi, Abdul Sattar, and a name I cannot recall. He is from the Albu Gharraf tribe. Sheikh Fara Sabar Deri from the Ghali tribe, and Sheikh Khalid Araq.

After the killing of Sheikh Khalid Araq al-Ataymi, who is from the Albu Aligasim tribe—they severed his head, they left him out in the open. His head may have never been recovered. We used that as a pretense to start the revolution. This guy was killed merely because he wanted to start a revolution, even though he had not even begun it. He wanted to start fighting terrorism. The name of his organization would be Anbar's Revolutionaries. We used his tribe under the pretext of vengeance. We elicited his tribe to help us. Their tribe also, because it is a part of Albu Aetha, wanted to do something, but they are not politically savvy. They do not have the political will, but we elicited their help nevertheless.

So Sheikh Sattar provided me with an office, with a computer, with papers to write on. His brother, Sheikh Ahmad, who was living in Dubai, wired me $5,000. At the time, Sheikh Ahmad was not known. And I started putting the group together for the revolution against al-Qaeda. Sheikh Ahmad was asking me to stay with Sheikh Sattar, because he did not have staying power, and if I left, he would be hanged.

The first people I tried to get were the religious clerics. I implored them that there was no sanctity left for travelers or for anybody else. Notable among them were Abdul Malik al-Saidi and Ahmed Al-Kubaisi. I exhorted them to Islamic principles. The second people I tried to get help from were the tribal sheikhs. Through the Internet, we got a fatwa to start the fight against terrorism.

The commander of the American forces at the time was constantly in touch with Sheikh Sattar Abu Risha, but I didn't meet them. The most important part was to finish off terrorism so that people would become free again, regardless of who we cooperated with.

So we started preparing for that conference we spoke of, and we let it be known that 90 percent of the tribal sheikhs were in Jordan. From over there, from abroad, we faced stiff resistance to the initiation of a counterterrorism undertaking. They accused us of trying to take over the tribes, trying to be sheikhs. They said, "You are not sheikhs. You cannot do this. You cannot do that." They mocked us. They said, "You're trying to take on al-Qaeda—the ones who fought America?" They laughed at us. They said, "You people are simple-minded. You cannot do what you think you're going to do."

Montgomery: Who told them this?

Interpreter: The sheikhs, the 90 percent of sheikhs that were living in Amman. Do you want me to ask him that? That's what he just told me.

Montgomery: No, if that's what he said. I just missed a word.

Interpreter: The sheikhs, the tribal sheikhs who were living in Amman, who were scolding them that he tried to take our place. They said, "How are you trying to fight al-Qaeda? And besides, you're not even real sheikhs. We are the sheikhs. You're trying to use our place. You're trying to pretend to be us."

Montgomery: No, I missed the word Amman.

Wissam: Even in Syria, even in Jordan and Egypt.

The Marine commander at the time pledged his support to the undertaking.

Montgomery: Does he remember his name?

Wissam: The 14th of September 2006, if you look at that date, you'll find his name.* There was a common interest between us and them, and that is security for all. I asked the commander at the time to help us by removing the concrete barriers, because they sectioned Ramadi into too many places. So he immediately said, "Yes, I will give you fuel, water, rations, remove the barriers." He said, "This would be a good step forward, the people will find out we are here to help them, not to harm them as the terrorists have done in the past."

What I'm speaking of at this time is the day before the conference. On the eve of that conference, a civilian vehicle came to Sheikh Abdul Sattar's. I think it was from the CIA [Central Intelligence Agency]. I was not able to meet with them. After the meeting, he informed me of all that went on because it was his habit to inform me of everything. He told me, "This is the speech that we have to give," and as I understood, the speech was from President [George W.] Bush. These are secrets that nobody is privy to except me and Sattar.

I was able to read the speech. Sheikh Sattar was not able to read it because he was illiterate. He did not finish school. And I told him, "If you read it at the conference, you will be laughed at, because this translation is verbatim, and they can tell this is not our writing. It is American writing." I changed quite a bit in the speech that we received from President Bush. The substance of the speech was left intact, but I changed the wording to grammatically correct Arabic that would be easily understood by the participants of my conference. Had it been left the way it was, there would have been a misunderstanding.

---

* Major General Richard C. Zilmer was commander of I Marine Expeditionary Force (Forward) at the time.

On 14 September 2006 we started.

Interpreter: And he did say a while ago, the speech contained something about the rebuilding of Iraq and pulling the troops out.

Wissam: We readied the place for the conference I'm speaking of. There were no television stations present at the time, so we brought a very simple camera to the conference. In that conference, we made a committee, which I personally called the Awakening and the Anbar Salvation Council. We chose 41 people to represent the governing council for Anbar, which was not present at the time. Sheikh Sattar was chosen as governor, and I was his administrative deputy. Sheikh Hamid al-Hayess was the head of the Salvation Council. Many more names are documented.

Here the political phase started. How do we implement the decision? The tribes began fighting, each one in its own territory. Each tribe knew who was a terrorist, so he was either killed, given to the government, to the police, banished, whatever. I admonished them not to kill as al-Qaeda was killing because if that was the case, then we'd never see the light of day; meaning give them to the law. Let the law take care of them. Most of the people in the Awakening did not have political savvy. They did not know the law. They did not have political awareness, and I warned them that should they kill anybody, they might end up in detention centers. [conversation with interpreter]

Interpreter: The Awakening started in Jazeera, in all these places he mentioned, except Albu Beid and Albu Bali. He mentioned Albu Aetha, Albu Diyad, Albu Faraj. Okay, Albu Faraj was the first to raise the surrender flag, so Albu Bali and Albu Baid—Albu Aetha, Albu Bali. Okay, so those are four places: Albu Aetha, Albu Beid, Albu Bali, and Garban.

Montgomery: And those are where the bad guys surrendered?

Interpreter: No, the bad guys surrendered in Albu Faraj.

Wissam: In Sufiyah and Street 17, the terrorists were entrenched. They were killing in the street. No man could raise his head. One day, when we were meeting, Sheikh Jassim Suwadawi from Albu

Soda—and there was another person with him, who came from Shamiyah—they came in, and they pledged their allegiance to the Awakening, and they asked for help. We gave him very simple assistance, a few rifles. Why? Because we were not helped by the army, by the government, or by the Americans. We were simply that involved, and we were told, "The enemy is behind you, and the sea is in front of you."

Interpreter: Meaning "between a rock and a hard place."

Wissam: The first cell was constituted to fight in Shamiyah with the leadership of Jassim [Muhammad Saleh] al-Suwadawi. I was called "the engineer of the Awakening." As far as Sheikh Jassim al-Suwadawi is concerned, believe me, he has many chapters. I lived them. He got involved in many battles with me. He had a great role, and he will never be given his just dues.

Now we come to the Americans' role. Now we come to the promises that were made by the American commander, as far as lifting the barriers, giving out fuel, kerosene, gasoline, what have you. We got in touch with Baghdad, specifically with Methal al-Alousi. We had no way of communicating with the prime minister. Methal al-Alousi was a friend of mine. We told him what we were undertaking. And he said, "You are heroes. Tomorrow I'm getting my bags, and I'm coming to join you."

We got a date to meet with the prime minister, [Nouri Kamil Mohammed Hasan al-] Maliki. The political entities started pressuring the American commanders in Anbar not to help us. We met with al-Maliki, and a verbal confrontation ensued, and we threatened him. If we were not given any assistance from them, we were going to push the terrorists into their territory, close the borders, and let them worry about the terrorists. Mowaffak al-Rubaie, the national security advisor, was writing, and he never said a word. But two days later he came on television and said, "Men of Anbar, we are supplying you with men and materiel."

When the people got wind of the fact that the government was about to help us out with men and materiel—meaning guns—they said immediately that we sold out to the government. Now we're lining our pockets, we're taking ammo and not giving it to the people. We're taking weapons and not giving them to the people, meaning the revolution is not genuine anymore, meaning his speech was full of landmines.

I got in touch with Methal al-Alousi. I said, "Okay, when are you coming? Bring your bags and come join us."

He said, "Look, we were told by President Bush, 'Whoever goes to the governor's office is to be shot.'"

I asked, "Why?" The political entities in Baghdad understood our Awakening to be against them, and that we were about to go into Baghdad and take care of them. So there was pressure from these political entities on the ambassador, and the ambassador on Bush, and back.

We came back, and the American commander at the time in Anbar had switched his thinking 180 degrees. I asked him about his pledge to help us. He said, "Look, I cannot help you. As far as Jazeera goes, I'm not going to help you with logistics, or weapons, or what have you." As for the governor's office, he said, "Do not advance against it."

Interpreter: I think I'm misunderstanding, because he said there is no help coming through for Jazeera. Do you want me to ask for that again, because he's saying now that "six months later we were able to liberate the governor's office in Ramadi."

Montgomery: Yes, make sure you're understanding.

Wissam: [Conversation with interpreter.] These were his instructions. You can go ahead and fight in al-Jazeera, but if we find any terrorist weapons, he would take them himself and not assist us with any kind of weapons.

The one thing that you have to understand is that the Americans were not sure of the loyalty of the Awakening people because the people that were working with the Americans told them, "These men are just wearing masks. They're going to turn against you, and they're going to kill you." It is a fact that some of the police were members of al-Qaeda.

Everything was confused in the meantime, except the determination of the Awakening people. Many problems ensued between the Awakening and the Islamic Party, so the American command suggested a meeting between us and them. The Islamic Party became wary of the Awakening. . . .

It's a long history, and it cannot be recalled in one hour. I went to al-Rashid. There was a conference there, and they said, "Here comes the engineer of the Awakening." So all the television reporters came and gathered around me because they were trying to make a documentary on the Awakening.

As I said, they suggested that we meet with the Islamic Party representative, along with the American command. Sa'doun Dulaimi, who was not at the Ministry of Defense at the time, advised me not to enter the meeting. I asked him why, and he said, "Because those are politicians, and they spin everything. You will never be able to advance an argument with them, because they will overwhelm you."

I told him there was no problem. We'd meet in Fallujah. The meeting was between me, Sattar Abu Risha, the governor, and his deputy. They took us from here via helicopter. . . . We didn't know what the future held. Are these guys going to dump us somewhere? I'm being frank. Sheikh Sattar was sitting right next to me. He said, "Look, any moment, these guys are going to push us out of the plane, because these guys like the governor." In the helicopter, with our garb, and the wind was flowing violently, so we were disheveled, but when we got to Fallujah, [Major General Richard C.] Zilmer greeted us. He had a complete team from the American embassy.

I had written down in a small booklet 24 points on Mamoun [Sami Rashid al-Alwani], who was the governor. Sattar asked me, "What am I to say?" I said, "Just start anything, and I will take over." That is because he was the head of the Awakening, and I was his deputy. He went in, and he said, "Greetings my brother, and this is Sheikh al-Hardan, and he will inform you about everything."

The discussion was bitter and sorrowful. When I confronted Mamoun with the 24 points I had laid out, he said, "Should I answer?" I said, "No, you answer one after the other." I said, "Write down one by one, and then answer." He said, "I don't have a pen." I said, "Your deputy has a pen." . . . When his deputy said, "I don't have a pen, either," I told Commander Zilmer, "Look, this is a very hot province, and this man is incapable of defending himself or carrying a pen. How can we leave him to be a governor?" It was the knockout punch to Mamoun. He came to me, and he pleaded with me to be his friend.

Zilmer suggested we form two committees to govern—five people from the Awakening and five people from the Islamic Party. After that, we kissed and made up, and we took pictures, and Zilmer went home. But then Mamoun abrogated the agreement. They started a division between me and Sheikh Sattar through his brother, Sheikh Ahmad, who was the closest to the Islamic Party, which led me to stay away from Sheikh Sattar for three months. From that day on, the Awakening was either infiltrated or sold—however you want to characterize it—and it ended with the killing of Sheikh Sattar.

These are the highlights, and there are a lot more. I was involved in a battle against al-Qaeda. The Americans helped me with it. So there is a lot more to be said, but I'm going to tell you when the film is over, so you're not recording.

Interview 5

Sheikh Jassim Muhammad Saleh al-Suwadawi

Contesting Sheikh of the Albu Soda Tribe
President of the Sufiya Awakening Council

Sheikh Abdul Rahman al-Janabi

Member of the Awakening

Sheikh Jassim Muhammad Saleh al-Suwadawi served as a warrant officer in the Iraqi air force for 30 years. He was an ordnance expert who specialized in armaments for MiG-23 and MiG-29 aircraft. After the true, lineal sheikh of the Albu Soda tribe fled the country, Jassim became the de facto head of the tribe, primarily because of his actions during and after the battle of Sufiyah.

Sheikh Abdul Rahman al-Janabi is a kinsman and a member of the adjacent Albu Mahal tribe. He served as a company-grade officer in Iraqi special forces during the Iran-Iraq War and was wounded twice. After leaving military service in 1991, he was recalled to the army and served as a company commander from 2001 to 2003.

Sheikh Jassim and Sheikh Abdul Rahman worked closely together before, during, and after the battle of Sufiyah. In the first interview, they provide an overview of events leading up to the battle, the battle generally, and subsequent developments. In the second interview, they give detailed descriptions of tactics and techniques employed during the battle and subsequent operations.

Estimates vary widely as to how many al-Qaeda fighters attacked the Albu Soda tribe on 25 November 2006. Coalition estimates are much lower than those of the men who were actually under fire, but all agree that Sheikh Jassim and Sheikh Abdul Rahman were heavily outnumbered.

Sheikh Jassim and Sheikh Abdul Rahman were interviewed by Colonel Gary W. Montgomery and Chief Warrant Officer-4 Timothy S. McWilliams on 12 and 20 February 2009 in the Albu Soda tribal area east of Ramadi city.

Sheikh Jassim Muhammad Saleh al-Suwadawi: After the disbanding of the Iraqi army and some other ministries, I had nothing to do. I had no job. I tried to help the Coalition forces, but it was difficult.

On 5 November 2003, I met a guy named Ayham al-Samarie to coordinate activities with the Coalition forces. He was the electricity minister. It was difficult to meet as frequently as I wanted due to the inherent danger of traveling to Baghdad and terrorism being rampant. I tried to elicit the help of the government of Iraq and Coalition forces through Ayham al-Samarie, and I tried to establish a covert chain of command. We gave periodic reports to Coalition forces and to the Iraqi government from November 2003 until September 2006.

On 28 September 2006, one of my brothers and three of my tribesmen were kidnapped. Then I declared open hostilities against the terrorists, and I established checkpoints in my territory.

I'm a member of the Tribes Council in Iraq. In September 2006, I had only 17 people fighting with me, so I got in touch with Sheikh Sattar Abu Risha and Hamid al-Hayess to get their help. I did not resort to them because I was defeated. No, I wanted to get permission to carry arms in my own territory. I met with General [Richard C.] Zilmer at the time. I met with Lieutenant Colonel [James] Lechner [USA]. I tried to get the help of Coalition forces. I tried to get better arms for my men, but there was somewhat of a mistrust in the Coalition forces to my aim because they thought it was foolish for 17 men to try and take on the entire terrorist apparatus around here.

The way I got through to Sheikh Abdul Sattar and Sheikh Hamid al-Hayess, I used to swim the river to get there safely, and come back by swimming the river. The distance was about 200 or 300 meters. It was very tiring, and one time I almost drowned, but, with the grace of God, I survived. I resorted to swimming the river because the place was infested with terrorists, even inside my tribe. They tried to kill me several times. They set up a lot of ambushes around the river to try to catch me and dispose of me.

The first thing I ever got from General Zilmer and General Murthi [Mishin Rafa Farahan], the commander of the Iraqi Army 7th Division, was a handwritten note saying it was okay for me to cross the river using a boat. . . . There was a combined American and Iraqi army checkpoint on the other side. That's why I wanted that authorization, so I could show it to them. That paper was dated 28 October 2006. . . .

On 20 November 2006, Coalition forces came in at night and arrested my men who were manning the checkpoints. There was no coordination between me and Battalion 1-9 of the Army. They didn't have prior knowledge that I had set up checkpoints to prevent the terrorists from lobbing mortars on them. My brothers and my cousins pleaded with the Americans that "we are here because Sheikh Abdul Sattar, Sheikh Hamid al-Hayess, and Sheikh Jassim put us in here. They need us in here to fight terrorists." They still handcuffed them and brought them in to meet with me.

Interpreter: He did say "with the permission of Lieutenant Colonel Lechner," too, that they were in there.

Jassim: The last thing I did, I sold my last cow to purchase a Thuraya phone to get in touch with Sheikh Abdul Sattar and Sheikh Hamid al-Hayess to coordinate with them. After they verified with Sheikh Abdul Sattar, Lieutenant Colonel Lechner, and Sheikh Hamid al-Hayess, they said, "Okay, you're okay." So they untied their hands and let them continue.

On the 24th of November, I negotiated with some of the chieftains in al-Qaeda to try to avert mayhem. It's because I squeezed them. I prevented them from accomplishing their missions because I set up the checkpoints, and I prevented them from going about in my territory. So they wanted to defuse the situation and try to convince me to let them do what they wanted to do. . . . They negotiated with me to lay down my arms, and they offered me as much as 1 billion Iraqi dinars.

At the time, they had already decided beforehand that if I couldn't turn an agreement with them right then and there, they would kidnap me and do away with me. They had prepared 17 vehicles loaded with men and weapons to kidnap me. The meeting took place at Sheikh Hamid Jabour's house. I wanted to delay the confrontation with them to a later time when I could better equip myself with men and materiel, but I kind of sensed they were up to no good. I had five armed men with me, and we pulled out of the meeting.

We wanted to renegotiate in a week so as to reassess our situation and to avert conflict. But at the same time, one of the Albu Mahal tribe was assassinated—which is adjacent to me—by al-Qaeda. The Albu Mahal are my cousins. They are my children's uncles. They are relatives.

I went back and reproached them. I said, "Look, you've opened another wound. You've just killed one of my tribesmen from Albu Mahal. We've just agreed that there will be no killing. Therefore, no agreement, no peace with you, period."

The next day, 25 November 2006, there was a huge attack by 850 armed men, armed with all kinds of weapons, from al-Qaeda and some other militia groups. They lobbed between 62 and 67 mortar rounds on my tribe that day. Afterward, they came in with their light machine guns. We fought back with the 17 men that I had. I suffered seven casualties, which I considered to be my brothers, but they were my tribesmen. And we killed more than 90 people from al-Qaeda. Among my tribesmen, I suffered 10 more casualties, consisting of women, older people, children—innocent people. The terrorists came into their houses and just slaughtered them.

The battle took place from 1:00 in the afternoon until 9:00 at night. We exhausted our ammunition. They were able to penetrate my house and my brother's house and burn them to the ground with all the furniture in them. They burned 11 vehicles that belonged to me and my brothers. At 10:00 at night, Coalition forces intervened, 1-9 Battalion.

We were able, with Coalition forces, with the Iraqi army, with my men—they were called Jassim's Militia at the time—to raid the places where the terrorists were concentrating. We were able to capture quite a few terrorists, and from interrogating them, we were able to get a lot more. Some of them I turned over to Coalition forces; some of them I turned over to the Iraqi justice system.

When Lieutenant Colonel [Charles] Ferry [USA] found out that I could confront the terrorists, he had enough trust in me, and I gave him prior knowledge as to where my target was. He started assisting me. Through that, I was able to clear the area little by little.

Colonel Gary W. Montgomery: When you're outnumbered so very much, how do you fight that many men? What tactics do you use? Were you at a house? What was the battle like?

Jassim: We used rooftops and ground emplacements, and we picked them off. We knew who they were because they were wearing masks. Our faces were open. We used maneuver to our advantage. Whenever we got pushed back, we pulled back. Whenever the force was overwhelming, we pulled back to other secure places until nighttime. And they were more busy picking up their wounded and dead comrades than trying to fight. By nighttime, they were finally convinced it was a lost battle. They were busy picking up their men because they didn't want to be identified. In addition, American jet fighters joined in. With the sound, I guess they got scared.

At 1:30 p.m. I called some security organizations. I told them there's a big fight in my area, but the American infantry did not join in, but the air force did join in late. After the terrorists pulled out, two of their vehicles were burned by the jets. . . .

I started an education program, trying to educate people to the fact that we were a small band of men who defeated a huge number of men. That is because God granted us the victory because we were in the right and the terrorists were in the wrong. So I was able to bring in many people on my side. Eastern Ramadi has about 12 tribes. Within three months, I was able to enlist many of these tribes into the police fold.

Sheikh Abdul Sattar asked me to gather the tribal sheikhs on 22 February, along with Colonel [John W.] Charlton [USA], and Abdullah [Muhammad] Badir, of the 1st [Iraqi] Division. I must try and praise him. There is a lot of praise for him because he participated in helping me establishing security in this area. . . .

In March of 2007, we started to change the tone of things from combat to providing infrastructure services to the people. So we started a council like the al-Jazeera council, and we nominated four people from each tribe. From 44, I was able to garner 41 in the council. Through that council, and with the help of Coalition forces, we were able to provide the people with many services. Through that initiative, we were able to provide the people with drinking water, with many services, and with jobs, because the area had been neglected since 2004. Through the Coalition forces, I was able to get enough projects going, such as pumping stations, paving roads, repairing and refurbishing schools, clinics, building schools. We were able to put people to work. We were able to put engineers to work . . . to repair the electricity and water. . . .

And now I will turn it over to Sheikh Abdul Rahman. He will give you the highlights. I really have done a disservice, because I didn't even mention his name, even though he really is one of the heroes that I depended on in the fight. . . .

Sheikh Abdul Rahman al-Janabi: On 17 September 2003, to be exact, Brigadier General Ibrahim Sijil Said, my uncle, got in touch with me. We used to work internal security in the former regime. Later, he was kidnapped by al-Qaeda, along with his son and my other cousin, and got killed. He wanted me to help him due to my extensive contacts with other tribes in the middle of Anbar, and in order to not let the Iranians penetrate into Anbar or Iraq as a whole. I agreed with him, and he told me there were some people in Baghdad who would support us. After he got their agreement, I went to Baghdad, and we met with them. (I don't want to mention who those people were.) We agreed to cooperate with them, and we formed a cell. We called it the Eagles Cell, and both I and my brother, Sheikh Jassim, belonged to it. From that point, we started our work, and we expanded it, providing the security apparatus with information about the terrorists. And that was 5 November 2003.

February 2004, the martyr Fatha, who also coordinated and cooperated with us; he used to carry my Thuraya phone because I couldn't bring it to Ramadi, and the GPS [global positioning system], he used to keep it for me, too. He was killed right in front of his house in 2004 in al-Dour by a terrorist squad. From then on, we worked with great caution and excluded any unit that had an Iraqi interpreter.

We still suspect that those interpreters are the ones who related information about us to the terrorists. One of the mistakes that Coalition forces made was using local interpreters instead of bringing interpreters from the United States. And we tried to alert them to the fact that those interpreters were duplicitous.

We worked until the ominous day of 29 April 2004. I call it ominous because my brother and three of his children were kidnapped that day. He was kidnapped on his way to Baghdad by a cell headed by Omar Hadid in Fallujah.

To tell you the truth, not only al-Qaeda fought us. All the armed militias fought us. They issued fatwas against us, and they sent out a lot of leaflets. That's why our work became very difficult and inherently dangerous for us and our families. Our support was minimal—you could say nonexistent—from our Coalition partners.

Now, not only were we supporting our families from our meager resources, but we were supporting the families of the martyrs who were killed supporting us. Coupled with that, we were not working our regular jobs, because the terrorists did not allow us to live like ordinary people. They fought us, and our resources, and our working areas. . . .

We attempted an uprising against al-Qaeda many times, and I will mention one. It was at the end of 2004, beginning of 2005. I met with the electricity minister of Ayad Allawi's government, Dr. Ayham al-Samarie. He mentioned a tribal uprising against al-Qaeda, and he said that he would provide me with support through Coalition forces in Baghdad. Indeed, we started that endeavor, Sheikh Jassim and me, and we made many contacts in Anbar and Baghdad, and Salah ad Din, Diyala and Kirkuk, in Baghdad, and Latifiya. We were trying to establish a public base to fight terrorism, but because of his frequent travels and lack of knowledge of the support people, and the change of government from Allawi to another government, that project was abandoned. . . .

All along, from 2003 until the uprising, terrorists were constantly picking at the people who got in touch with us, and, as we say, "It reached up to here." When they kidnapped my uncle, Abdul Sattar Sijil—that was the 18th of October, 2004—that's when my fight began with them officially. With the help of Sheikh Jassim Muhammed, I succeeded in saving my uncle from their hands. We wounded one of the people who kidnapped my uncle.

In July or August 2006, they kidnapped my brother. I rescued him and found him between life and death due to the extensive torture that he suffered. After all of that was done to us, Sheikh Jassim and I started gathering information relating to these terrorists and their places, their emplacements, and their arms, until the day that I call the ominous day, when his brother—

Interpreter: He wants to say something. The brother of Sheikh Jassim was kidnapped on that day.

Jassim: There was a general consensus among the people in the street that those people were resistance fighters, so we started a campaign of education that these people, if they are truly resistance fighters, why are they kidnapping our men? So during that, we were able to garner a lot more information about the terrorists.

Abdul Rahman: In that operation, Sheikh Jassim Muhammad had only a pistol, but he was able to save his brother. He fought them with his pistol and wounded one, but they got away.

On 20 November 2006, that's when Sheikh Jassim made the courageous stand in open warfare against them and placing checkpoints all over the place—him, his brothers and his children—despite the objections of some of the tribal chiefs in the area. Those sheikhs were in cahoots with the terrorists.

This movement that Sheikh Jassim started, started the Awakening in eastern Ramadi and the tribes of Ramadi, along to the Abu Risha sector, along to al-Fallujah from the side of Ramadi. . . . He got in touch with Sheikh Abu Risha, God bless him, and he officially provided him with support, and Sheikh Abdul Sattar was in the forefront in giving the blessing for this initiative. He and the tribal sheikhs called him the Lion of Eastern Ramadi, and well deserved.

At the time, al-Qaeda started planning an offensive against the sheikh's tribe. They perpetrated many assassinations of people that were close to him, but he remained steadfast despite all the obstacles and challenges. He spent all of his resources in support of that initiative. He did it so people could live in security and safety.

Again, the offensive began on 25 November 2006. At 1:30 in the afternoon, the criminals started their offensive with probably more than 850 fighters. They started by lobbing more than 67 mortar rounds. But the courageous fighters, despite their low numbers, were able to withstand and thwart the offensive.

At the time, I was at the funeral of a sheikh from the Albu Mahal tribe, Salman Adnan Hallawi. When the fight started, . . . my uncle from Albu Mahal tribe came in support. Only six men came with him from the Albu Mahal tribe. I was one of them. Two of them were martyred. And from the fighters that Sheikh Jassim had, there were 17 fighters, and some of them were martyred. From the information we gathered after the battle, the enemy suffered about 90 dead and 170 wounded.

Right then and there, the barrier of fear was broken in all of Ramadi, so open warfare against the terrorists took place. We had a great role in clearing the areas I'm about to mention—Sheikh Jassim and me. The sector from eastern Ramadi to Khalidiyah, al-Jazeera, Albu Beid; al-Jazeera, Abu Ali; wal-Karma; Ibrahim Ibn Ali, which is located east of Karma toward Baghdad. Our fight against the terrorists continued up until now, and right now, we are fighting them in seven provinces.

Second Interview

Jassim: In order to protect my tribesmen from outside terrorists and from internal terrorism within my tribe, I had 17 fighters. I placed points at the entrances of the tribe. Their main objective was to search vehicles coming in and going out.

Two simple attacks took place upon the people that were manning these checkpoints. They lobbed some mortar rounds at the tribal area where we live. This took place about a month before the main battle. And we had skirmishes with small arms between the members of my tribe and the terrorists.

We had small Motorola radios, and that's how we communicated. I had four checkpoints. Each had one radio. At no given time was there less than two to four people manning each checkpoint. They changed shifts day and night. My men carried AK-47s [assault rifles]. Each one had 30 to 40 rounds. One PKC [machine gun]. So for the 17 fighters I had, I had 16 AK-47s and one PKC. They were very vigilant all the time. . . .

The attack began, and they entered the area from three different points. There was no equality in the forces. We had 17 fighters, while they had about 850 fighters on their side. Even in the weaponry, there was no parity, because 850 fighters would carry a lot more than 17, and ammunition as well.

After sunset, some of my men started pulling out from their positions because they ran out of ammunition. Because of my men pulling out, the terrorists were able to penetrate the area. They started killing children, old men, and women. They burned two houses that belonged to me. They killed cattle and sheep. They burned vehicles. Even electric generators—they started shooting at them, and burning them, and destroying them.

I had a Thuraya phone, and I was getting in touch with my men in the field and the American forces as well. The battery ran out in the Thuraya. Coalition forces were able to locate exactly where the battle was taking place through the GPS [global positioning system] in the Thuraya phone. They sent fighter planes to fly over the area. The Coalition forces were asking me for exact locations of the enemy, but when the enemy penetrated my area, everything was intermingled. In military terms, when everything is intermingled, there is no specific target, so the battle is lost. So I told them, if they want to hit the terrorists when they pull back, start hitting them, so we would inflict heavier losses on the enemy than on ours. And indeed, they were able to burn three vehicles after they pulled out, with all the occupants in them, about 15 kilometers from here.

We assessed our losses after 9:00, after the battle. We found that we had 17 martyrs from the men, children, and women, and seven of my fighters were also killed. Those seven were missing, I should say. We couldn't find their bodies. They had tied up their corpses, and they were pulling them behind their vehicles for about 20 kilometers. Five of them were my brothers. They were dragging them.

We found many of the enemies' corpses on the battlefield, even though they had taken many with them. Some of their corpses stayed on the battlefield for about three days. We tried to exchange corpses with the enemy, but it did not take place.

The Iraqi army and the 1-9 [battalion, USA] came into the field at 10:00 at night. They set up positions within the houses of my tribe, so we started raiding the enemy's hiding places, day and night. . . .

Because of my military background—I had 30 years in the military and had studied psychology—I was supervising the interrogations myself. I would send a raiding party for five people. My men would bring me 30. Believe it or not, I put the people in a room like this, the 30 of them. In addition to the information that I had already gathered on them, I knew who was a terrorist and who was not just by studying the expression on their faces. I was accurate most of the time. So the people I let go were honorable people, and they carried arms with me, not against me.

What I used to do with them, I used to gather them in a room like this, and I would make a speech about the evil of terrorism and the destructive nature of it, and after that, they would believe in my cause. After two or three days, they would come back to ask me to let them become policemen. Within a very short period of time, we were able to organize 12 tribes against terrorism, despite the knowledge there were some sleeper cells entering the tribes. . . .

Abdul Rahman: I want to give you an abbreviated synopsis of what happened to the plan. The sheikh, when he laid out the plan in advance, he did not imagine the amount of force that was coming against him. His plan was a defensive one, for a very small force. . . . But because of his dogged determination, with the help of the Almighty, the whole, full-fledged battle took place on his hands.

Some of the men who sympathized with him but could not help him, they helped him a little, either directly or indirectly. I'll give you an example. When al-Qaeda came to attack, they did not expect the Albu Mahal tribe to resist them at all. They thought they would have safe passage, but the Albu Mahal attacked them and sniped at them.

So the plan, as it was set up, was defensive in nature; but as the battle developed, it was spontaneous. The battle itself imposed on us what we had to do. Initially, he relied on his own resources and his cousins' resources, but as the battle developed, he started asking for help from Abu Risha, from Coalition forces, from the Iraqi army.

Montgomery: We say that no plan survives contact with the enemy. You always have to adjust.

Abdul Rahman: Exactly. After the battle, we considered three months to be the golden months, because the battle developed square-by-square, and it was to our advantage. First, we were able to win the people over against al-Qaeda. Through the good people, we were able to educate the people about terrorism and against it. We encouraged the opening of police stations so the law can take hold. Despite the fact that the youth were reluctant to join the police because of fear, our education and our lectures got them into the fold, and we were able to open the first police station. It was named Karameh, which meant "dignity."

After that, smaller battles took place all throughout al-Jazeera. They saw the example of what we had achieved with small means. They rose up, and they helped clear al-Jazeera. I got in touch with Mohiad, Ibrahim, and Ahmed, and through the media I was able to announce that this tribe had joined us and that tribe had joined us. Despite the fact that maybe one or two of the tribe came with me, I said the whole tribe was with me. So people starting enlisting on our side, because they're saying, "Look, the whole tribe's with them." That way, I won them over, and I was able to outwit the al-Qaeda through the media.

After the battle, we did not just lay back and say, "Okay, it's over." We kept gathering intelligence on their cells. We found many a weapons cache, because the terrorists had relied on them—like you see a fighter coming from Diyala to here, his weapons are already here, but we had already confiscated them.

We used the media to our advantage. If we had a 10 percent success, we said it's 90 percent.

Montgomery: Right. Information operations.

Abdul Rahman: Yes. So, you see, when a tribal sheikh started talking about a success, even if only 2 percent was true and 98 percent was false, well, when the terrorist heard it, it's grand, it's big. So they started to flee the area by themselves.

We used the counterattack. Instead of waiting for the enemy to attack again, we pressed the attack on them. When they got confessions from detainees as to the existence of other places, despite the fact that they have no communication gear, no vehicles to travel with, they would conduct four raids a night on these targets.

There was a well-known saying that Sheikh Jassim was going to get a chopping machine that would chop the terrorists to pieces, so everybody knew to avoid Sheikh Jassim. The truth of the matter, everyone was dealt with fairly, either through the Albu Risha, Coalition, or Iraqi army, the government, what have you. But the general consensus was Sheikh Jassim was chopping people up—just to let the terrorists get wind of it and instill fear in their hearts.

After the battle, the sheikh formed a cell to gather intelligence and another unit to raid the terrorists. At the same time, he extended his hand to the central government and established law and order in his territory. He told them, "Look, I am a tribal sheikh, but I cannot be the police. I cannot be the interrogator. I cannot be everybody. Let's divide the jobs, one for each."

After the success of this battle, many a tribesman came in from al-Jazeera and from the surrounding areas, trying to learn from the sheikh's experiences.

One of the misinformation campaigns that I conducted, I would pledge that a person is good. I'd sign it, and put my seal on it, and give it to his people, saying, "Look, there is nothing on your man or on your son. Let him come back." And when he came back, we detained him.

I would like to relay one incident that took place. The terrorists came into this man's house, who's from the Albu Mahal tribe, and they took his rifle so he could not shoot at them. After the battle, we went out in the sector that I was in. We found a dead terrorist, among many, who had the man's rifle next to him. So we took the rifle, and we gave it to its rightful owner.

Jassim: The three months that followed the battle, we were surrounded. We were under an embargo. We could get no medicine, no food, no water, no fuel, and it was a very severe winter. If they got wind of anyone supplying us with fuel, food, anything to sustain life, they would kill him. With my meager means, I bought three boats. We started buying things to sustain life, fuel, food, medicine for the sick and all of that, from al-Jazeera, through the three boats, because Jazeera was more peaceful, more stable than us. . . .

Whenever I wanted to go on a mission, say, a raiding party, I used misinformation. Like, I'd tell Coalition forces that I'm going to the Albu Alwan area on a raiding party at night, and I asked them for lights so they can identify my men. That night, I changed the plan. Why? Because these phones are easily infiltrated, and they would know my plan ahead of time, so I used misinformation. I kept them guessing at what I was doing. And I changed the plan right when we were working toward the objective, so that nobody knew there was a change of plan, and without informing Coalition forces of my change of plan.

Montgomery: Wasn't that dangerous?

Jassim: Yes.

Montgomery: Did Coalition forces ever shoot at you because you weren't where they thought you'd be?

Jassim: No, no, because we had the chemical lights.

Interpreter: He's talking about the chem lights that the Coalition forces supplied him with. So that was their signal. Because they carried them, they knew who they were.

Montgomery: That's what I thought.

Jassim: Twice I had raids on terrorists, and I did not find the targets, so I knew the Motorolas were infiltrated. It could be infiltrated easily. That's what made me think, what if those people have it? . . .

So at times, when I told the Coalition forces I'm going here, I changed my plan and went there. And that's where I found the targets. Not here, but there.

One time Lieutenant Colonel Ferry asked me why I changed plans. I told him why: "Because I think these radios are infiltrated. The terrorists could easily listen to the Motorolas." Indeed, 10 of these Motorolas were found with the terrorists. Some of them had the Iraqi army signal, and some of them had the Coalition signal. . . .

I was able to enlist the help of one of the people of the Albu Fahd tribe. I gave him a Thuraya telephone, and he would call me and tell me where the terrorists were, and we'd raid them. He used to come from his bedroom, and you know the Thuraya telephone is not very effective from inside buildings, so I used to hear his voice breaking up, but still, through his talk I was able to know exactly where the terrorists were located.

I set up a task force, no more than 30 fighters. I obtained some chem lights from the Coalition forces. We used to used this road right here, which is called Seda Road. I was in constant communication with my men from the beginning of the raid until coming back. I'd give them the exact location of the target, and within an hour, they'd call me and say how many men they got and how many weapons.

The Coalition forces wanted to join me in many raids, and I refused. Why? Because Coalition forces, when they go, they use vehicles. Vehicles make noise. Noise alerts the enemy to run away, so I prefer to go dismounted. So I preferred the Coalition forces to be at least 300 meters away from my raiding party, as a backup.

And the roads we took to get these people were infested with IEDs [improvised explosive devices]. These IEDs were rigged with wires and a battery to initiate an explosion to kill my men. Because of my military background, working in the air force, I was able to show my men how to disarm these IEDs by cutting the wires. They used to bring me terrorists, and along with them they used to bring IEDs on their shoulders. . . . The terrorists surrounded themselves with IEDs.

Montgomery: On the roads?

Jassim: On the roads, right. . . .

Some of the terrorists were very close to the encampments of the Americans. Why? Many a terrorist camped right here [pointing on a map to a neighborhood adjacent to an American compound], close to the Coalition forces. Why? These areas right here, nobody can stay in them except with the permission of Coalition forces. So what we did, we got the Coalition forces' permission to get into this area close by them, and we found many terrorists.

This is also a camp of the Coalition forces [pointing to another place on the map]. These houses right here [points to adjacent neighborhood], we found many a terrorist. They have many ways of hiding—misinformation—you cannot count them. . . .

Montgomery: Did you have to go at night?

Jassim: Day and night.

Montgomery: Did there seem to be more in one place, or about the same number coming from all the directions? Was there like a main attack and supporting attacks, or was it equal all the way around?

Jassim: Let me explain to you. [He points to various locations on a map as he describes the disposition of forces and scheme of maneuver.]. . . . All of them wore face masks, and they were in succession. The guys that wore red head bands were on suicide missions. Then yellow bands, and then black, and their clothes were black. There were terrorists coming from Hit, from Baghdad, from Fallujah, from Diyala. They all joined in this attack, and it was a huge attack. Why did they concentrate so much in this attack? They theorized that if I succeeded in clearing this area right here, all the surrounding areas would follow suit, so they wanted to finish me off right here. But thanks to the almighty, their thinking was faulty.

Montgomery: When they attacked, were they on line or were they moving like we do, in groups that are in formations, or was it sort of disorganized? What did it look like?

Jassim: The attack took stages. Like I said, those terrorists who came in on a suicide mission, when they achieved an objective, others came in and took their places, and so forth and so on.

Interpreter: It's not exactly the answer to your question.

Montgomery: No, but it helps. . . . Did you have anyone carrying ammunition up to the checkpoints before they ran out?

Jassim: No, I didn't have any men resupplying my men. Everyone relied on his 30 or 40 rounds. I was not able to buy ammo. Nobody ever gave me any ammo. People with ammo were afraid to sell me any, because terrorists were killing them.

Montgomery: What about your wounded? Did you have wounded, and what did you do with them?

Jassim: Yes, Coalition forces were able to treat my wounded and took some of them all the way to Balad, and I thank them for it.

Montgomery: You faced a very highly coordinated attack.

Jassim: Yes indeed. It was a military plan, and it had plenty of intelligence, and I tell you, some of the intelligence came from my own tribe. They got the exact location, who was carrying what in the checkpoint, and the numbers of my men. They had already known beforehand they did not have a whole tribe fighting them, just 15 or 17 men against all of them, so they imagined it was very easy to finish us off, and quickly. We were able to capture or detain some of the men who were relaying information to them and who let the terrorists stay in their own houses prior to the attack. They are still being detained. They confessed to everything they did, and hopefully they will get their due from the justice system.

Montgomery: Did you have any thoughts of trying to fall back across the river?

Jassim: Some of my own relatives fell back and crossed the river.

Interpreter: And I asked him why your relatives?

Jassim: Because they were fearful of being slain and their heads cut off. These men did not have any weapons, and they were afraid that, because they were my relatives, the enemy would capture them and chop off their heads. Some of my relatives—men, women and children—they had no weapons, but they killed them just because they were my relatives. This cousin right here, they killed his sister and they chopped off her head, just because she was my sister, and they cut off her breasts. . . .

Unknown: Did he have enough time, advance knowledge of the attack, to do any defensive preparations? And if he did, what did he do to prepare?

Jassim: I didn't have any. No, I had no prior knowledge as to the exact time of attack.

Unknown: Could you ask him, of the terrorists, how many were foreign, how many were Iraqi?

Jassim: Due to the fact that we found no identification papers on them, we did not know who they were, but I estimate there were about 50 foreign Arabs. We found some corpses that, from the looks of them, were not Iraqis.

Montgomery: Then when you brought in the captured prisoners from these raids, for interrogation, you said you looked at them, and a lot of times you could tell with good accuracy who was an insurgent and who wasn't. What were you looking for?

Jassim: When I gathered these men—and based on the intelligence that I'd already gathered—when I asked one of them a specific question, I saw his hands shake a little bit, or he couldn't look directly in my eyes. I knew, and he was afraid. These were some of the signs that I used to identify the terrorists from the nonterrorists. I used to look at their features, everything in their faces—even their ears. And I would ask him a question—how far in education did he go? If he had no education, I knew he was influenced by the terrorists. The uneducated were easily swayed to be on the terrorist side, and so I would concentrate on him and ask him more questions because they used ignorant people, uneducated people.

Interview 6

Sheikh Aifan Sadun al-Issawi

Fallujah Representative
Iraqi Awakening Political Party

Sheikh Aifan Sadun al-Issawi's father opposed the Iraqi nationalists of the 1960s because they were interfering with family and religious traditions. In 1967, he killed a man at a meeting in Fallujah and fled with his family to Saudi Arabia, where King Faisal provided him with land, homes, and support. He subsequently died during a visit to Helwan, Egypt, and his family believes that he was poisoned by agents of the Iraqi government.

Sheikh Aifan was born in 1972 in Ar Ar, Saudi Arabia, on the border with Iraq. He studied computers and mathematics in college and worked in Saudi Arabia until 2001, when he moved to Iraq to take care of his mother after her second husband died. Nine of his brothers still reside in Saudi Arabia, and two live in the United States.

Sheikh Aifan is the nephew of Sheikh Khamis Hasnawi al-Issawi, the paramount sheikh of the Albu Issa tribe. Sheikh Khamis exercises leadership of the tribe, but he is aged and not very active. Consequently, Sheikh Aifan tends to represent the tribe in dealings with Coalition forces.

Sheikh Aifan was interviewed by Colonel Gary W. Montgomery and Chief Warrant Officer-4 Timothy S. McWilliams on 14 February 2009 at Camp Ramadi.

Sheikh Aifan Sadun al-Issawi: As we feared, there was a war. We believed there would be a war, so as a normal precaution, we all armed ourselves and prepared for it. I was almost certain the Iranians would enter Iraq.

The collision came quickly. Baghdad was finished, but Fallujah was still under control. We contacted Abdul Dureb, the director of security in Fallujah. He asked me to visit him, so I went to his office. I found about 50 people there. They were foreign Arabs. I remember one of them. I knew him. He's from the Emirates. There were Syrians and Tunisians—a lot of people. Some were Fedayeen Saddam. I heard Abdul Dureb tell the Fedayeen Saddam, who were wearing black uniforms, to take these young people [the foreign Arabs] and prepare them to fight. He told me that we all had to fight. I told him that Iraq was finished, to just stand by because fighting was useless. After that, I left.

A few days later, I heard that he had disappeared, and the Coalition forces had arrived at K-160 [highway kilometer marker 160]. The tribal sheikhs had a meeting in our guesthouse. As I recall, they had many ideas, but in the end, we decided to let the Coalition forces enter the cities. We made an invitation to meet the American force at K-160. We negotiated with them. We even had lunch together in the desert. We said, "If you want to go inside the city, come on in, but without fighting." We had a deal, and they came to Ramadi and Fallujah, and the surrounding cities like Habbaniyah. They made a temporary camp at the big bridge by Fallujah, but they didn't enter the city.

I stopped to talk with them. They had an interpreter, an Egyptian named Ennis. And there was a captain named Hickey. They were my points of contact. I told them I would give them advice on how to behave when they went into the city. I also told them about an Iraqi army base in front of Taqqadum, Habbaniyah. It was full of weapons. Any weapons you can think of, it was full of them. It was so big you needed a car to go around it. It was huge. Captain Hickey told me, "We'll go together, and I'll see it." We went in two helicopters, and I showed them the location. They were surprised. When we came back, they told me they would send a force to surround this camp, but they didn't do anything after that.

After the Americans went into Fallujah, people started going into the bases and taking weapons because they were not secured. I took about 80 RPGs [rocket-propelled grenades], PKCs [machine guns], and other things.

The Americans had no experience with our culture. The British, when they went to Basrah, they knew the area, because they've been there. They knew how to deal with the sheikhs and with the people. They had ways of communicating with the sheikhs and the people.

But the Americans, when they got inside Fallujah, they started to arrest people. They arrested sheikhs.

They shot me by mistake when we had a meeting in Karma. They shot me in the left leg. My tribe took me to our hospital. You know that I am a son of the sheikh of the Albu Issa tribe, so many people know me. Everyone was surrounding the hospital to find out what's happening to me. And it was at the beginning. It was a new thing. So when they heard that the Americans shot someone, everyone was surprised. In our culture, they go to find out what's happening.

Then the Americans visited me in the hospital and said they would take me to Germany for treatment. We have our habits and our traditions, and it was too great a favor. The Americans were a new force in our lives, and it seemed strange, so I refused their offer. They went to the government people in Fallujah to get them to make me accept their offer, because the wound was a bad one. The bones were broken. So they put some Iraqi security people around me in the hospital to protect me. I was bleeding, and it was a very, very dangerous situation. They gave me blood transfusions to stabilize me so they could fly me to Germany. The Americans visited me in the hospital and said that if they didn't take me to Germany, my leg would have to be amputated.

Some of the sons of Albu Issa, my tribe, were very brave people. They took me from the hospital by force. My friends, my sons, my people—they know me in the tribe. They surrounded the hospital. They went to the policemen who were protecting me and locked them in a room. Then they took me to Amman, Jordan. My brothers came from Saudi Arabia to Jordan, and they were waiting at the border. They put me in Khalidi Hospital, which is the best hospital in Amman. They took care of me over there, and they operated on my leg. They replaced the demolished bone with a platinum tube. They found that my blood had been poisoned, and they thought the bullet that shot me contained some kind of poison. I stayed there about four months. After six months, I came back to Iraq. I was on crutches. I had a very hard time.

So I arrived home from Amman one night, and the second night, the Americans surrounded the area where my house is. They even had helicopters over my house. They arrested me and my brothers—Sheikh Barakat, Sheikh Jamal, and their sons. So the whole group was about nine. They took us to Fallujah town.

General King* was responsible for the campaign; his headquarters was in Baghdad, but he was responsible for this campaign [conversation with interpreter]. He was upset with me because, maybe, first because I refused their offer and went to Jordan. And he was upset because Barakat, my brother, refused to shake hands with him.

Then they took us to Abu Ghraib jail. I stayed there nine months without being charged with anything. We didn't know the charges, and I was very sick. There were demonstrations in Fallujah to release me, a delegation of sheikhs interceded for my release, and many things happened like that. But any time they asked General King to release me, he refused very strongly.

Then Jalal Talabani, the Iraqi president, asked President [George W.] Bush to release me. So they came to Abu Ghraib jail, and they took us by helicopter. They took us to the Iraqi president, Jalal Talabani. We stayed in his palace as his guests for about three days. Then they put us under house arrest for about six months.

Colonel Gary W. Montgomery: Could you describe what it was like being imprisoned in Abu Ghraib?

Aifan: It was like a grave. I wore the same clothes for nine months, and I was very sick, and it was often very cold. We didn't know what was happening outside. There were always mortars falling on the jail. They put us in a tent, and many of my friends inside the tent were killed by mortars. There were foreign Arabs with us. Abu Ghraib jail was very bad. They scared us with attack dogs during interrogations.

And this lady—you know the pictures?** When the thing happened in Abu Ghraib? This lady with the short hair? I know her personally, and she knows me personally. I was one of her prisoners.

---

* Possibly then-Col R. Alan King, USA.

** Probably Pvt Lynndie R. England, USAR, who was shown in many of the infamous photographs from the prison.

But when some of the Americans came and learned that I can speak English and saw me on my crutches, they sympathized with me. They took my detainee number, and after one or two days, they came back and said, "You have a big, big problem. We know you are a good man, but we can't help you. We need to help you, but it's a big problem." [inaudible] I had the crutches, so [inaudible]. They allowed me to go to my house in Fallujah, on one condition—that I won't leave Iraq.

During the time we were in Abu Ghraib, the situation was getting bad in Fallujah. Fighting started to increase between the Americans and some fighters in Fallujah. So when we were with the president of Iraq, Jalal Talabani, he talked with us in the presence of some of the Americans. They said, "We're going to release you to cool things down in Fallujah." So it's a deal. When I came back to Fallujah, the situation was very critical, and the insurgents were rising up.

After we went home, the Americans visited me in my home. There was no one being visited by the Americans in his home. It was strange, because the insurgents, if they had any suspicion that a guy had any relationship with the Americans, they cut off his head. The leaders of the insurgents in Fallujah at that time were Abdullah al-Janabi and Omar al-Hadid. These were the top leaders of the fighters there. There were a lot of foreign Arabs with them. After the Americans visited me, the insurgents posted a lot of papers around saying that I was an agent for the Americans.

Then I was meeting the Americans at the center called FLT [Fallujah Liaison Team], which is from the American and the [inaudible] hospital in Fallujah, which is on the highway.

Montgomery: Is that the small compound just east of the Jordanian hospital?

Aifan: Yes. It had some old rooms that had been used for the Iraqi army, but I rehabilitated it and made some offices for the Americans, so they could use it as an office for reconstruction or compensation for the people. People were coming to ask about their compensation. If someone was hit by the Americans, or his car was burned by the Americans, he came, and the Americans paid him for it. And the Americans, for history and to be honest, they helped the people too much.

The head of it was Major Gregory [G.] Gillette. He is a lawyer who was responsible for the Civil-Military Operations Center. And there was Colonel [Colin P.] McNease, Colonel [John A.] Toolan [Jr.]—he's a general now. He was the top leader of the forces in Fallujah. He is a very close friend, a very dear friend. McNease and Gillette were very close friends to me.

So I was going to this place daily. I helped people get compensation, and I was still on crutches. Then the Americans did the paperwork for me to get compensation for my leg. They asked for the invoices from Khalidi Hospital in Amman. They saw how much it cost me, and they paid me back.

I started to develop a stronger relationship with the Americans, so the insurgents made me a target. They couldn't hit me, because I have my own people who protect me, so they couldn't face me face to face. But they kept me in mind as a target, watching for me. We kept each other afraid of each other. Then they hit one of my cars, and they killed one of my people.

So I called a huge meeting in Fallujah. I collected all the people—the imams of the mosques, and the sheikhs, and everyone high level in Fallujah. I have a videotape of the meeting.

Montgomery: Do you remember what time this was, approximate date?

Aifan: I have the tape, and I have the date and the time.

Montgomery: Was it between the battles of Fallujah, or was it before the fighting started?

Aifan: Before the big fighting in Fallujah. There were some people who were insurgents or mujahideen, but they were not against me. They attended this meeting. There were armed people around all the roads. They went all over the area. I held this meeting in coordination with the Coalition forces. I let the imams start to talk and tell the people that the insurgents who come from outside Iraq and cover their faces, they are not mujahideen. They are criminals. And they warned people that they were letting surrounding countries, like Iran, Saudi Arabia, and Syria, interfere inside Iraq.

Sheikh Hamza [Abbas al-Issawi, Mufti of Fallujah], God bless him, was the head of the imams in Fallujah. He attended that meeting. After that, he was killed. They killed him in front of the mosque. Sheikh Jamal [Shakir al-Nazzal] and Sheikh Kamal [Shakir al-Nazzal] were the top imams in Fallujah. Sheikh Kamal was killed.* Sheikh Jamal fled to Syria, and he's still there.

After this meeting, the fight between us was announced. We started to coordinate and work with the Coalition forces, and we started to target the leaders of al-Qaeda. We killed some of the big leaders.

I started fighting al-Qaeda publicly in 2005. During this time, from my cousins, my families, and my tribe's sons, about 37 were killed because of this fighting. In 2006, after the big battle in Fallujah, the situation was getting very bad, and people were going after to the foreign Arabs. People wanted to live peacefully, so they started to follow al-Qaeda. There were just seven people being faithful and staying with me. I was surrounded, so I couldn't even leave my house. If I wanted to go outside, it meant too much security, and I might have to fight at any second. So I started to not go out much. Even my uncles, who are just across the road from my house, told me, "Don't come to us, because you are a suspect. They might kill us because we know you."

Montgomery: What year is this, you're talking about?

Aifan: The end of 2005.

Montgomery: Okay.

Aifan: At the start of 2006, they ambushed one of my cars. We tried to visit Sattar Abu Risha. No one could visit Sattar Abu Risha at that time, because whoever visited him would be slaughtered.

I had a good relationship with Sheikh Sattar and Sheikh [Hamid] Jabour. So I went to Sheikh Sattar's house, God bless him. I found just two people with him. We approached his house and waited on the highway, going back and forth until the highway was completely empty.

---

* Sheikh Hamza was assassinated on 29 November 2005, and Sheikh Kamal met the same end on 7 February 2006.

Interpreter: Because his guesthouse is very close to the highway, less than 100 meters.

Aifan: We left Sattar's house after lunch. We took the old Habbaniyah road to Fallujah. Al-Qaeda was at the Boston and Ira intersection.* There were about seven cars waiting for us, and we only had three cars. We faced them and started fighting. Of my people, four were killed and one was very seriously shot. We killed one of them and seriously wounded three.

The four who were killed had been in one car, which was disabled, and we had killed one of them [al-Qaeda]. We kept going until we reached home with our wounded man. The situation was very bad.

I took my family to Amman, Jordan. I have my own house in Amman, so I stayed with my family—my wife and my children—in Amman. And even Sheikh Sattar fled to Amman for a little while. He visited me, and we talked about the situation, saying "what are we going to do?" Then Sheikh Khamis [Hasnawi al-Issawi, paramount sheikh of the Albu Issa tribe] and Sheikh Khalid, my uncles, fled to Amman. They refused to fight al-Qaeda. They came to Amman and lived with me in my house. I stayed about four months in Amman.

The Americans visited me in Amman. General [David G.] Reist—and there was colonel—

Interpreter: There was a colonel who was responsible. He doesn't remember his name.

Aifan: He visited me in Fallujah, and it was very bad. We are still in 2006. He wanted me to come back and fight al-Qaeda.

Sattar came back, and he started fighting al-Qaeda. Then Sattar called me and said, "Hey, look, I'm in Ramadi, you are in Fallujah. Let's cooperate and start fighting al-Qaeda and finish them." The biggest problem that had kept us from fighting al-Qaeda was that we didn't have weapons. The Americans at that time took any weapons they found, anywhere in any house. So I made a deal with

* Boston and Ira are the Coalition names for these two roads.

the Americans. I will come back to Fallujah from Amman, but keep away from my weapons. We had a deal with them. So that's why I came back and started fighting.

The first day I came back, I was attacked by mortars. You know the news in Iraq—everybody knew that Aifan was back, so they started shooting at me with mortars. The Americans made a brave stand. They stayed with me at my house for a long time. There were many attacks on my house, and they were on the roof. They stayed with me. We fought together on the roof of the house.

So al-Qaeda started to expand their shooting. They started shooting all the houses around me, and it was an advantage for me—expanding the fighting, so it's not just me. Because if I stay alone, and I fight alone, and they're targeting me alone, they will hit me. I would be a very easy target. So it's better for me to expand it and make it bigger.

And the other advantage, if I expand the battle, there will be too many people with me. So we expanded the fighting for the Albu Yusuf, Albu Jamali area, Albu Khalid—around my house, around my area.

Then al-Qaeda sent a letter to Sheikh Khamis, my uncle. The letter was delivered by Sheikh Abdullah Shuesh, one of the Albu Issa tribal sheikhs. The letter said, "We want to meet Sheikh Khalid and stop the fighting." So Sheikh Khalid met them. They said, "We don't have problems with you, the Albu Issa tribe. We have a problem with Aifan, and Aifan is an agent. Either you kill him, and the problem will be finished, or bring him to us, and the problem with you will be finished."

After this situation, my cousins sympathized with me. Some of my cousins joined me, and they are heroes. I had eight PKCs. The Americans helped me with some ammunition, so I started getting stronger. The first operation I did was in a center for al-Qaeda, and I did it with the help of the police.

I had three black BMWs, and I was always changing cars so nobody knows which one I'm riding in—three of them, same type, same color, everything. So they put some IEDs [improvised explosive devices] on the road. One of my cars was blown up. With me was Mushtar, my uncle Khalid, Othman, and my uncle Majid. Mushtar lost both legs, Othman lost one leg, and the driver was killed.

The Americans helped. They took them to al-Asad by helicopter. And then I took them to Amman to take care of them, to Germany, and now they are back, but without hands and without legs.

After they were wounded, their families turned against me, but I kept fighting. Al-Qaeda sent a tanker of chlorine against me. They blew it up in the first checkpoint, which is very close to my house. My mother was killed—her name was Turkiyah—and five children were killed, and one of my guards. This gave me another great push to fight them. I still kept my arms on me, and we joined the American operations to fight al-Qaeda. And we kept fighting until God, by his blessing, gave us the victory.

In 2006, the Awakening was announced publicly. I think I'm a person who was close to Sattar and Ahmad Abu Risha, and they made a very big sacrifice. The Americans supported us very strongly through the projects they put in our areas, and also through the police and the Iraqi army, which we started with them and the Americans trained.

In 2006, I established an ERU [emergency response unit] battalion in Fallujah, and I was the leader. It is a very strong, very well-trained battalion, and it protects Fallujah. And thanks for that.

In my house, I have hundreds of pictures. If we spent three days or a week just talking about writing, you would not understand exactly what was going on as well as if you saw the pictures, the videotape. We filmed our operations. We documented everything. I have pictures of the Iranian we caught in 2006. Two people who tried to kill me in 2007, I have a picture of them. Three weeks ago, they sent a guy with a vest.

Interpreter: Three weeks ago?

Aifan: To blow himself up. And I still have one guy, a foreign Arab. He's Yemeni. I have him in my jail now. You know, we can't just kill him. If I kill him now, they will be happy, but other people will come after a couple of years and take me to court and punish me because I killed him. It's happened. We have law. We should send him to the court and the law, and if the court decides to kill him, they can kill him in the street.

Interpreter: I told him, "Sir, if somebody came to kill me, I would slaughter him, I would kill him."

Aifan: When we were fighting with PKCs, face to face, we killed many people.

Interpreter: Yes, in fighting.

Aifan: Yes, one from Sudan, one from Saudi Arabia. Me, I killed them.

Interpreter: During battles, yes.

Aifan: But if we catch someone, and we know we have rules and our religion, we can't kill him. A Syrian, if he killed my mother, and I catch the guy, I can't kill him.

Interpreter: In our religion, if you catch someone, it's not good to kill him. He's helpless. He's hopeless. So for that, he said, he's had for two months now a guy that has been arrested, and he can't.

Aifan: Eleven times they tried to kill me. Last year, they killed my captain and two of my guards. They came to my checkpoint and said they were British priests. They came to my area, and they had American uniforms and a British badge. He was Tunisian, I think. His face looked American, and he made himself American. He said, "We need to talk to Sheikh Aifan." But I went to Jordan that morning, and he came to my gate at 1:00 in the afternoon. When they told him I was outside Iraq, he went crazy. They were sure I was at home. He said, "I need to talk to whoever is the chief of security around Sheikh Aifan. Who is it? I need to talk to him." Captain Mahmoud was my security guy. When he came to him, he caught him like this, and he blew himself up.

There was a woman with pistols. When they came to detain her, she blew herself up.

Montgomery: You said he made himself look American.

Aifan: Yes.

Montgomery: Do you know where he was from?

Aifan: He was Tunisian.

Montgomery: Tunisia, Okay.

Aifan: The Americans know about that. They did an investigation. He was from Tunisia, and he had a fake police badge.

Interpreter: And he looked white, with blue eyes.

Aifan: Yes. A week before that, they sent two kids with vests to the mosque on a Friday, and they were waiting for me to enter the mosque to blow themselves up. But we ambushed them behind the mosque and caught them. We called the Americans. They came and took the TNT and all of that stuff, and we sent them to jail. They are in jail now.

Montgomery: Do you think they're after you now for political reasons, or out of revenge?

Aifan: Before, they tried to kill me because we fought al-Qaeda, and I am sure they will keep looking to kill me. I have heard the Americans say, "He is a target for the rest of his life," and they gave me letters for the American government saying that "he did this and that, and he helped us. We think he deserves to be an American citizen, he and his family, because he will remain a target for the rest of his life—him and his kids."

I have four kids. I didn't put them in school. They will kill them, and you know they will kill them, especially now.

When we fought al-Qaeda, we didn't have a big problem, because we fought al-Qaeda, and 1920 [Revolution Brigade] fought al-Qaeda, the Islamic Party fought al-Qaeda, the Islamic army, the mujahideen army—all these groups fought al-Qaeda the same as me, so that we destroyed al-Qaeda completely, on the ground and on the table. Underground, we still have sleeper cells. On the ground, under your eyes, we destroyed al-Qaeda.

Now we face a big problem. Some of those groups who fought al-Qaeda with us had a different agenda. Yeah, they fought al-Qaeda with us, because we had the same enemy. But after they finished al-Qaeda, they started looking to kill us, because we are "agents" who work with the Americans.

Me and my people, we didn't fight al-Qaeda because we have religion or ideology or something. We fought al-Qaeda because they are bad people. They killed people and did evil things, so we fought them. I don't care if they pray or don't pray, if they believe in God, or have some other reason. I don't care. But the Islamic Party, the Islamic army, the mujahideen army—all these groups, 1920—they have ideologies, different agendas. Some of those groups have foreign agendas, maybe from Iran or Syria.

Now we're a bigger target, more than when we fought al-Qaeda, because now I am sure 1920 are bad people. They tried to kill my people. And the Islamic Party, I know they mean to kill me. I am sure. We have proved it.

Interpreter: Before, you saw them. Now, you don't see them.

Aifan: Now, some of the people who were in those groups are in the government. He is captain or major or something, so that he has the ability to move very freely, and they can watch me, and they can listen to my radio. So it's very, very difficult.

Interpreter: More risky now.

Aifan: I am sure they will keep on targeting me because I am not in one of those groups. And they'd fight with me. And they always visit me.

Interpreter: To join them?

Aifan: Yes, they told me, "We will make you a big head," and "Come with us." "You are the sheikh."

Yes, I am the son of the sheikh. But everything ends—everything. The soldiers, the leaders, the mayor, the governor, the minister, [Prime Minister Nouri al-] Maliki, Jalal Talabani—for some reason, they will all be changed. But I will still, for all of my life, be the son of the sheikh. I am a good man. I have my respect and my people. I think my level is very high—the highest.

Interpreter: Of course, they're going to lose their positions, but your position is for life.

Aifan: If I am poor, I am the same Sheikh Aifan. If I am rich, then I am still Sheikh Aifan.

Montgomery: So you don't need them, but they need you.

Aifan: Yes, sure. We are still a big target. I can't move, except with my guards, with my armored vehicle. I have very high security around my house. It's difficult.

I have a Saudi passport. I have a Jordanian passport.

Interpreter: I told him, "You have a passport, why don't you leave?" He said, "What about my history? This is my history, my life."

Aifan: My people, my tribe. I can't just leave.

Interpreter: He has to be faithful for them. It's easy for him; he has his passport.

Montgomery: What would happen to your tribe if you do leave? Would a brother take your place here?

Aifan: No, I don't think so. We should be honest. Life won't stop if I leave, but I have a thousand people who trust me. They follow me blindly. A thousand people. Leave Iraq now?

Interpreter: It's not faithful to leave them.

Aifan: It would be a black point in my history if I leave them. We have to be willing participate more in the government. We have to have balance in the government—the government of Iraq and Anbar—because these people's rights are my responsibility. I should take care of them.

God's will made me honored. I am the son of the sheikh, and I am the leader, and I am the youngest one in the family. I have 16 brothers, and 14 brothers are older than me, but God chose me. And the people chose me. They came to my house. They didn't go to my brothers' houses. So if you see, if you know my life and how hard it is, and how the people came to me and asked for my help, you can't just leave them. It would be unfaithful to leave them and run away.

Montgomery: I think with so many brothers, and so many brothers who are older than you, I guess it's not like in Europe with the nobility, where the oldest brother gets the title. How are you chosen from among your brothers?

Aifan: They chose me. Nine brothers are Saudi, and they were already in Saudi Arabia. Three of my brothers have no interest in the tribes. They are businessmen, and they are not ready to spend one hour listening to the people. And Barakat, who is the oldest, I have not agreed with him for the last five years. He is very hard, and he's against me being a friend of the Americans or making any relationship with them. He was against the fight against al-Qaeda. He lives in Jordan. He was just back for a couple of months. So the people look for the guy who guided them, who showed them, who made something a fact—not some big shot, but someone who helped them.

I am not strong through my muscles; I am strong through my people. When I make a celebration in Fallujah, thousands of people attend. I think you saw the celebration for the Awakening that was on TV. It was the biggest celebration in al-Anbar by attendance, because people love me, and I am very strong through them. I am not strong through my money, or the Americans, or through my muscles. When I am upset, I stand up, and hundreds of people stand up with me. That's what I'm working for, and I'm ready to sacrifice all my life for these people.

Montgomery: I understand better now, and it sounds to me like it's not even something like a formal election so much as the people just sort of begin to follow the person whom they trust most and have the most confidence in, among the brothers. They are drawn to you.

Aifan: And because I'm the son of Sadun, and I was the closest one to my father in his behavior. I was very similar to my father. When they saw me, how I behaved, they said you look like your father. I'm very proud of that.

Montgomery: What was it like in Fallujah between the two battles? And what was the effect of the battle of Fallujah?

Aifan: They hated the Americans. It was a very big mistake. They could have taken care of it a better way. From the beginning, if they dealt with the right people, they wouldn't need to fight by themselves here.

Montgomery: Since we've been here, since 2003, what are the biggest mistakes we've made, and what have we done right?

Interpreter: In Fallujah particularly, or in Iraq, sir?

Montgomery: In his opinion, and in any regard—wide open question.

Aifan: The biggest mistake was the incident in Abu Ghraib. After that, there was something—American soldiers raped an Iraqi woman. I think you remember this story. Even if the Americans punished him, and I am sure the main government didn't agree with these mistakes. But sometimes you have to be responsible for what you've done.

And the other mistake, which made these mistakes go on and on, was that the Americans dealt with the wrong people. They didn't mean to do it, but they believed—because they are honest, they believed anyone—so they dealt with the wrong people. It's not the Americans' mistake.

We can't—I should be honest—we can't trust the Americans completely yet. We have a theory. The Americans don't have continuous friendships. They always have their interests. Their relationships with people are based on how much benefit they can get from a person. When we were fighting al-Qaeda, the Americans were in my house every two days. I even met President Bush. [Barak H.] Obama—I saw him; [David H.] Petraeus [USA], all the generals.

Each MEF [Marine expeditionary force] changed. The generals invited me to celebrations, and we made relations stronger between us. The situation got better in Fallujah. When General [Walter E.]

Gaskin [Sr.] and General [John R.] Allen left, General [John F.] Kelly came in. Kelly visited me one time in my house, and General [Martin] Post, I didn't see him except in a general celebration. Now they are gone, and new generals are coming, and I don't know them.

This is a big mistake. It means the advantage of having a relationship, a friend, is finished for me. We have a good situation. It's safe. But a new person could say, "That's it—we don't need him anymore." I still have very good relationships with some of the Americans, mainly the leaders and generals. I should be in the meeting if they came, and they are leaving. At least call and tell me. This is a mistake.

Even I—and I'm very close to the Americans—am starting to be distanced from them. The government of Iraq, Nouri al-Maliki called me the "American sheikh." They know I have very strong relationships with the Americans, which makes me happy. I don't feel ashamed of it. I used my friendship with the Americans to build a secure Iraq.

This is what we have achieved, because it's God's wish, and with his blessing and the work between us and the Americans. Now I'm involved with the police, and we're taking care of the security situation in Fallujah and the surrounding area. And we are cutting out this cancer, al-Qaeda and the insurgents. And we won the election, thank God, and I am a provincial council member. Maybe I will be more than a provincial council member in the future. Maybe.

And we have deals on the table. Sheikh Ahmad [Abu Risha] was in Baghdad yesterday, and we have a meeting every day with someone on [inaudible] politics. We are working. Maybe we can get something bigger on the table. We keep going, working to build and secure our Iraq.

And a big mistake, which they will regret—the Americans pull out of the checkpoints very fast. This is a killer mistake. We have a lot of police, with good equipment and weapons, but they are compromised. We have some bad people among them. They change their faces.

Montgomery: Among those who are manning the checkpoints now?

Aifan: In the police. We detained some people who planted IEDs against the Americans. We caught them in the act. When they were detained by the police, they stayed in jail two days, and they released them, because they were placing IEDs against Americans, not against Iraqis.

And I'm very sure and very confident that there are policemen using police vehicles to place IEDs. And I'm very sure that the insurgents and al-Qaeda have keys to the checkpoints, and they can get through without any questions. This is a big mistake. But the Americans do it very fast.

Maybe it's Obama. I told him, when I talked to him. I gave him a letter. If you become president, be alert. Don't pull out very fast. I personally talked to him more than 20 minutes, but he is still young. He's going on with it.

Montgomery: During the fighting against al-Qaeda, was the Awakening a single organization, or several organizations that were fighting at the same time? Throughout Anbar, was there coordination between the anti-al-Qaeda fighters in each city across the province?

Aifan: There were many people that were fighting. There was the Islamic Party, 1920, [Interpreter and Aifan simultaneously naming various groups].

Some people from al-Qaeda changed their faces. They are in the Awakening now. They didn't join us because they believe we are good, but to protect themselves. They saw who's winning, so they said they were with the Awakening. But when al-Qaeda was in, they were al-Qaeda.

Montgomery: There are always people like that.

Aifan: People without any principles, like these people, you cannot trust them. The proof of it is the bad people infiltrating the police. And the proof of that is the incident when they killed Sheikh Abdul Sattar. The guy who planned it and did it was in his security detail, and he was captured.

Chief Warrant Office-4 Timothy S. McWilliams: The colonel asked about the period between the battles of Fallujah, from April 2004 and November 2004. Could you tell me what was going on in Fallujah during that time?

Aifan: After the first battle of Fallujah, they established a security force. They were all faithful to al-Qaeda. They were al-Qaeda, and they proved it. After that happened, there were no police on the street; just the people who wear masks and their drivers, which is al-Qaeda. It was a very big mistake.

I told Colonel Toolan and Colonel McNease, "One of these days, the Fallujah people will fight you with these weapons that you gave to the police," and that happened. They used police vehicles and the guns from the Americans to fight the Americans.

The Americans didn't listen to advice. The Americans don't trust anyone. They have the principle, "don't trust anybody, so you can protect yourself." They came to my house and slept in my house. We fought shoulder to shoulder. I have a long history with the Americans. And yet when I visited the FLT every Monday, they searched me.

One time I remember, at the center of operations in our area, they spent eight days in my house. They brought the screens and the satellite, the big chairs, and everything. We spent eight days in my house, like my brothers.

Interpreter: And they're still searching you.

McWilliams: You mentioned Abdullah Janabi and Omar Hadid, and I've heard they've done some terrible things. Can you describe some of those?

Aifan: They put evil thoughts into the minds of the people of Fallujah. They established the Shura council in Fallujah. They gave authorization to kill anyone who worked with the Americans and anyone who volunteered to be a policeman. And they were thieves. Abdullah Janabi owned buildings and factories in Syria. Omar Hadid was an outcast. In the Saddam [Hussein] regime, he was accused, and charged, and wanted in many sex crimes.

Who were the people who joined al-Qaeda? Violent and irresponsible people. He had no value in society. His mother or his sister was a prostitute. He's homosexual. He found a chance to control the people, and he was being humiliated by the people, so he joined al-Qaeda. So al-Qaeda are bullies. They were all named with their mothers' names, not with their fathers'. Did you find any sheikhs, any high-level people, who were involved with al-Qaeda? No way. Only outcasts joined al-Qaeda.

The most important thing is that we woke up from this bad dream.

Interview 7

Sheikh Ali Hatim Abd al-Razzaq Ali al-Sulayman al-Assafi

Heir Apparent to the Paramount Sheikh of the Dulaimi Tribal Confederation
and
Head of the National Salvation Front Political Party

Born in 1971, Sheikh Ali Hatim Abd al-Razzaq Ali al-Sulayman al-Assafi is relatively young, which is typical of leaders of the Awakening movement. He is also the true lineal heir of Sheikh Ali al-Sulayman, who was one of the most powerful sheikhs in al-Anbar during and after the transition from Ottoman to British rule (described in the writings of Gertrude Bell).

The role of paramount sheikh is nominally filled by his great uncles, Sheikh Amer and Sheikh Majed, who serve as co-regents. The former resides in Ramadi and the latter supports from Amman, Jordan. Sheikh Ali Hatim is the de facto paramount sheikh, though he declines to claim the title publicly out of respect for his uncles.

After the assassination of Sheikh Abdul Sattar Abu Risha, the Awakening movement fractured and Sheikh Ali Hatim formed his own political party.

Sheikh Ali Hatim was interviewed by Colonel Gary W. Montgomery and Chief Warrant Officer-4 Timothy S. McWilliams on 16 February 2009 at Camp Ramadi.

Sheikh Ali Hatim Abd al-Razzaq Ali al-Sulayman al-Assafi: We went through a difficult period that rattled us during the Saddam [Hussein] era. Before Saddam, we knew Ali Sulayman and two more sheikhs, and that was it [referring to his lineage]. During Saddam, we had 460 sheikhs. This would be considered interference in our independence.

The second difficult period that we experienced was the uprising of the sheikhs opposing al-Qaeda. I want to emphasize the fact that we did not bring al-Qaeda into Anbar. There were mistakes on both sides, on the tribal side and on the American side. There was a gap between the American forces and the tribes. There was no meeting of the minds, so to speak.

In 2004 al-Qaeda began to appear, and we warned General [Charles H.] Swannack [Jr., USA] at the time. Their presence after 2004 intensified, and if you look in the al-Qaeda ranks, you will find a lot of lowlifes. Their aim was to destroy the Islamic religion by attacking the Christians and Shiites, not just the Anbari sheikhs.

In 2005, we met with the defense minister and the interior minister at the time, and we asked them to institute a force to protect the western desert. We proposed to them that we had 2,500 fighters, and we were willing, able, and ready to attack al-Qaeda. We got permission to institute this force, but we found cowardice in our own men. I asked why, and they said because they were intimidated by al-Qaeda and its brutality.

At the beginning of 2006, the Albu Mahal tribe in al-Qaim, that was the first spark, between Albu Mahal and al-Qaeda. They instituted a company called al-Hamza Company, but they got no help, neither from the Americans nor from the Iraqi government. Al-Qaeda started buying people, buying their consciences, and this is what we warned about. The reason being there was a lot of unemployment, so they exploited that factor. We exploited the spark that took place in al-Qaim. We started educating the people here and tried to incite them to stand up to al-Qaeda. We succeeded in taking some people from Anbar to Baghdad to demand support from the government, because support was almost nil. We were followed by al-Qaeda, and the first attack occurred in a place called Sadiyah, in Baghdad.

Interpreter: I asked him what time that attack was on his person in Sadiyah.

Ali Hatim: In the beginning of 2006. And before that, they attacked me right here in my guesthouse. They totally wiped it out. Sheikh Amer [Abd al-Jabbar Ali al-Sulayman al-Assafi] was at the guesthouse at the time. Nevertheless, we insisted on finishing the endeavor that we had undertaken.

About mid-2006, we had a conference, which the American ambassador [Zalmay M. Khalilzad] attended, and some other notables. We gathered the most important sheikhs, even from the south of Iraq, and we proposed raising arms against any militias, whether Iranian-backed or al-Qaeda. Very few reacted positively, because they were fearful.

I tell you that right now, there are some new people who rose up that were unknown before. There was a movement called Anbar's Revolutionaries, which started three months prior to the Awakening. The first confrontation that they undertook was in the Rahman Mosque, right here in Ramadi. They detained 12 people, nine of whom were emirs of al-Qaeda, and they executed them. The second incident was in Haditha. They detained two people. They brought them to the Tash area and executed them.

We got in touch with Abdul Aziz al-Kabasi, the intelligence chief in Baghdad, and we asked him for a copying machine so that we could make leaflets to incite people to rise up against al-Qaeda. To be truthful with you, we tried to stay away from working with the Americans because we didn't want it to look like an American initiative, because people were opposed to that. We wanted to show that it was wholly Iraqi and to incite them and to bring them into the fold.

Montgomery: About what time period was this?

Ali Hatim: About two months before the inception of the Awakening. That was with the agreement of the American command in Baghdad. We got the American commander in Baghdad to agree to the wording that we were using, because the situation was not really solid here. It was probably a secret endeavor that we were undertaking.

The second spark, like I spoke of al-Qaim and the Albu Mahal tribe, the second spark, there was an officer who was traveling along the highway. His wife was the daughter of Hamid al-Odda, who was the sheikh of the Albu Ali Jassim tribe. Al-Qaeda was questioning him, and they got hold of his wife's hair and pulled her out of the vehicle.

We exploited that, and that was the second main spark that took place. We used that incident, of pulling the woman by her hair, to give the people passion, saying, "Arise! Look! This is your cowardice! This is your complacency! This is your ignorance! This is what it has done! Look what we have descended to!" So we poured gasoline on the fire, so to speak.

Colonel Gary W. Montgomery: Where was the incident?

Ali Hatim: Close to Jarash, which is here in Jazeera, here in Ramadi. The officer's name was Hamid Zivin.

The true campaign that took place against al-Qaeda, to be honest with you, was started by Hamid al-Hayess, long before Sattar Abu Risha. Right at Hamid al-Hayess's guesthouse, that's where the meetings took place and the incitement to fight off al-Qaeda. Hamid al-Hayess got hurt the most by al-Qaeda. That was the step that we took forward in calling the revolution against al-Qaeda. We used what we call our right, which is tribal right, to fight al-Qaeda. So Sheikh Hamid al-Hayess asked the tribal sheikhs to gather, to incite people, to bring people into the fold to fight off al-Qaeda. There was no safe place to go, except Sheikh Abdul Sattar Abu Risha's house. And we made many decisions that were very dangerous, and they were dangerous to implement. That is because al-Qaeda possessed lethal weapons, and we lacked support from anybody, Americans or the government.

And you know something? When we started talking about this initiative, many of the sheikhs thought we were a little bit mad, because here we are trying to take on al-Qaeda. They thought we were crazy, and most of them ran away. And you know something? I'm willing to confront them to their faces and show them their cowardly stance. Right now, when they meet with the Americans, they boast about how courageous they were, but the fact of the matter is that their own people know who they were and their cowardly stance.

So when we decided to have that meeting I talked about a while ago, we got in touch with Sattar Abu Risha, and I said, "You are welcome. My house is yours." So the Awakening revolution—now we're into the Awakening phase—was started by Sattar Abu Risha, Hamid al-Hayess, and me, and we also have some heroes. Notable among them is Mohayed Maish, Fathal Maklef, and there was a brigadier general. His name was Hamid Ibrahim Jazza. Sorry to tell you, he was detained by the Americans. Americans killed many al-Qaeda members. Another person that was with us, his name was Shalan al-Nouri. Some other men had the heart to join with us. And open warfare against al-Qaeda started. Now, thanks be to the Almighty, it started.

Afterward, we turned our attention to Nouri al-Maliki, the prime minister. Hamid al-Hayess and I and some other men, we met with al-Maliki, and it was a very long meeting that lasted from 10:00 in the morning to 6:00 in the evening. When we went, we asked him for many things. We asked him to create emergency response units, a police force, bring in the army, and without hesitation he did. He had a long meeting with his security staff, and they fulfilled all our demands.

The Awakening was basically a conference, and we decided, okay, after this conference, what is it we're going to do? What is the end result of this? So the end result was the Anbar Salvation Council. I'm going to speak frankly. The Anbar Salvation Council was to be in the forefront in the fight against al-Qaeda, and to lead Anbar Province, and to have people from that council appointed to the governorship, and to lead the government. At the time, there was no provincial council. At the time, the Islamic Party had run away from the province. It was like a vacuum. It is well known that the Islamic Party used the vacuum that it had created to take over the government in Anbar. People started coming to our side, seeing the government of Iraq was helping us, and seeing what we accomplished. I'll tell you something, the American side was important, even though it was a little bit late.

We had a conference in Baghdad to nominate someone who would lead the Anbar Salvation Council, and we nominated Hamid al-Hayess and Sheikh Sattar Abu Risha. Hamid won by 70 points. Sattar Abu Risha pulled out, and he said, "Look, I'm going back with the Awakening," and here the division took place. What we started was a great step, but what took place was the division that we had really feared. I don't know what it was, maybe a different point of view or whatever, but there was a difference.

Montgomery: About what date was that?

Ali Hatim: At the end of 2006, beginning of 2007, right about there.

Montgomery: You were attacked twice, once in Baghdad and once in your guesthouse. Could you describe the attacks?

Ali Hatim: I was traveling from Jadriyah to Sadiyah, visiting my in-law, and we had four vehicles. Once we made a right turn into Sadiyah, hell broke loose, fire from everywhere. I had four men killed and many more injured. Most of the vehicles were disabled, except one. I don't know how I was saved, except by the grace of God, because it was a one-sided confrontation. They had six vehicles. They had PKCs. They had heavier machine guns. I don't know how I was saved.

The second confrontation, after we came from Baghdad, we started going about in Shamiyah and Ramadi proper and holding education sessions with the people. One day, we're meeting—we tried to incite them to fight, not intelligence gathering. We were not really military commanders who knew what to do really. All that we wanted to do was fight, fight, fight. One day we're meeting with my people in my guesthouse, and a young kid came in. He was about 15½ years old, at about 7:30 in the evening, and he stood right in front of me and started pushing a button, like he was drugged. You could see the wire coming right here from his cuffs, so he started pushing, pushing, and nothing happened. I guess it malfunctioned. So my men came in and took him out to the ditches and killed him. And the American forces came in the next day and detonated his explosive payload.

But we insisted on finishing the undertaking we had started. We had accomplished something great. We had contacts with the Iraqi government, but to be truthful with you, between us and the Americans, there was a gap. That was during the time of the general who preceded [Major General John F.] Kelly.* To tell you the truth,

* The time frame before MGen Kelly arrived was February 2007 to February 2008, during which time MGen Walter E. Gaskin Sr. was Multi National Force-West commander.

we did not concentrate on the gap between us, because we were really busy doing what we had to do. We did what we could to strengthen the security apparatus in Anbar, to strengthen the institutions in Anbar.

We had some confrontations with the Islamic Party. We started demanding to know the Anbar budget. We tried to find out where the money went from the budget of 2006 and 2007. The money belongs to the Anbaris. They are the rightful owners of that money. And I tell you frankly, we insisted on kicking out the Islamic Party, even if we had to resort to the use of force.

The American command, to be truthful with you, before Kelly, was exasperating the situation, not making it better, and we gave up on it. It almost came to a confrontation between us and that general I spoke of. It is not really a good thing to bring a military man, who only wants to assert his dominance through the use of force. You have to bring someone who has the political savvy, the knowledge of the people, the tribal inner workings of the people, because Anbar is a tribal place.

We got an invitation from [Mahmoud Dawud al-] Mashhadani, who was the Iraqi head of the parliament. He asked us to be patient and not to have any conflict with the Islamic Party. At the same time, there was a change of the American command in Anbar. That's when General Kelly came in and took charge, him and General [Martin] Post. We really did not know who they were. When we met with Mashhadani, the head of the Iraqi parliament, we threatened them that within a month, if the Islamic Party was not removed, it would be subject to confrontation, maybe attack.

The first meeting between us and Generals Kelly and Post took place in Habbaniyah. I'll tell you something, if there are generals like Kelly, kind and important—General Kelly and General Post worked with us using our mentality, not the Marines' mentality. There was a consensus that there was a huge difference in the command between the one that preceded Kelly. And Kelly himself listened to us, tried to understand the problem, and solved it. Maliki wanted to see us, and we met with al-Maliki to solve this dispute. Kelly invited us for a lunch, and he asked us to be patient and said, "You have my support for anything you face." We found that Anbar Province was open to all: the prime minister, the Iraqi parliament, and the American command. For that, we met, we thought, and he helped us in many things, like how to solve our problems before we use force. So after the confrontation with al-Qaeda, the confrontation with the Islamic Party was the most notable, and I'll tell you, General Kelly was able to defuse it.

You know what really makes us mad? It is seeing that the Americans were able to spend money and fulfill projects much more than the Iraqi government, or I should say the Anbari government that's headed by the Islamic Party. What the Americans have done is great. What they have been able to accomplish is great, but also the flip side of it is, "What if the Americans had the Anbari government's money?" They would have accomplished a lot more.

President [George W.] Bush's visit to Anbar was an important thing as a political message, but not in substance. Nothing really came out of it. [Barak H.] Obama's visit to Anbar, along with the two senators, the Democratic senator and the Republican one, was an important one.* We met with them, and we gave them a clear message. The true test right now for the Americans is to leave Iraq standing on its own feet, not lying on its back.

And to tell you the truth, rockets, scientific things like genes and the science of genes, and stuff like that, are not as important as leaving Iraq standing on its feet. The true test for America is to leave a new Iraq, as we were promised. I'll tell you what, the true test of democracy came in this election, not the one preceding it, nothing before that, ever.

There are three things to note that were of very great importance. The first is the appearance of al-Qaeda in Iraq. It was something that was not imagined. The second thing was that the Americans

* Then-Senator Obama, who was at the time the presumptive Democratic Party nominee for president, visited Ramadi on 22 July 2008 as part of a congressional delegation that also included Senator John F. "Jack" Reed, a Democrat from Rhode Island, and Senator Charles T. "Chuck" Hagel, a Republican from Nebraska.

stood with us, how they backed us up. That was not imagined. And the third thing was Maliki's stand, and how he did not look at himself as being Shiite. He stood with us, regardless of his sect.

So there were three noted things that were of great importance that one must look at, after which we had two meetings in preparation for these last elections. I went to the council in Baghdad, and two VBIEDs [vehicle borne improvised explosive devices] came in and attacked us. They brought down the building on us and, thanks be to the Almighty, I was saved.

General Hickman* paid me a visit, the deputy ambassador paid me a visit, and many more people from the American command and the embassy came in and paid me a visit. Even at those precarious times, we still insisted that Iraq would be clean and clear of al-Qaeda.

I think the reason I was attacked was because, two weeks prior to the attack, I had made a statement on the air that the trouble and the problems were neither the Americans nor the Iraqi government, but it was the Iranians. I also said that there is no connection between the Iranians and the Shiites of Iraq. So, with the help of the Almighty, we are on this steadfastly. The rebuilding of Iraq, getting Iraq rid of all the militias. I espouse no chair, no position, nothing, just the service of Iraq.

Montgomery: I'd like to go back and ask two questions about what you said earlier. First, at one point you were facing al-Qaeda, who was armed with lethal weapons, but you could not go to the Americans because then it would look like you were their puppets. So how could you fight under those circumstances? What do you do to fight? It sounds like you're organizing, but how do you fight?

Ali Hatim: It's not the fact that the people would think we are American stooges. No. There was a huge mistrust between us and the Americans due to the fact—if you remember the attack on Fallujah?

Montgomery: Yes.

Ali Hatim: There was a huge gap between us and the Americans.

---

* Possibly Col William B. Hickman, USA.

That's why we didn't want to ask them. As far as your questions, "How did we fight off al-Qaeda," we were able to buy weapons. Everything was on the market. To buy a PKC [machine gun] was like 500 bucks. We were able to buy French weapons. We were able to buy heavy armaments, Dushka—a Russian Dushka antiaircraft [heavy machine gun]. So there was plenty of it going around. And due to the distrust that existed between us and the Americans, had we asked the Americans for weapons, they would not have given them to us because of the fear factor. They did not know who we were.

Just to let you know, it's a known fact that Iraq is awash with weaponry. That's one thing.

And the second thing that we counted on, the fact that the Americans would come on our side once we started the fight—and this is exactly what took place. To tell you the truth, if you look at the Americans' role, and what they did in here, they were more interested in the welfare of the Iraqi people than the government of Anbar was. Tell you some facts—the way they stood behind us, the way they supported us, the way they kept overwatch on the police apparatus, on the security apparatus. One thing about it, they even counted the gasoline, how to disperse it among the police, who took it in turn and sold it. So the Americans kept watch over everybody and anybody. They really kept everybody in line, and if it weren't for them, it was chaos.

Montgomery: You're leading into my second question. You said there was a change in the American approach when there was a change in command. Could you give some examples of how things were changed, before and after?

Ali Hatim: The difference between the two commands, the first guy was sort of moody. He tried to please some people, very few people, mostly the people that were outside, by giving them contracts that really didn't matter too much.

Montgomery: Outside where?

Ali Hatim: Outside the country, to the sheikhs who lived maybe in Syria or Emirates or Jordan. So he was trying to please these people, rather than to do the right thing in Anbar, like I said. He was moody. He didn't have the political savvy. He did not know how to work the Anbar people.

Whereas General Kelly, no, no, no—he started using reason with us. He helped us. He did everything possible to defuse the situation against the Islamic Party because, I'll tell you what, if the confrontation had taken place between us and the Islamic Party, not even 100 American tanks would have defused it.

We had intended to finish off the Islamic Party. The thing we detest the most is Islamic Parties in Iraq. I'll tell you a synopsis of how he treated Anbar. He treated it like his own garden. He paid a great deal of attention to the cowards who ran away from here, the people who stayed in Jordan or Syria, or what have you. He gave armored vehicles to some people—some people who did not deserve it. He put people who did not deserve to be in leadership ahead of the real leaders. They purposely wanted to insult, dishonor, and create disputes between people. So he really did not—this is the first commander.

Montgomery: Isn't that what Saddam did, created new sheikhs?

Ali Hatim: Yes.

Montgomery: So it's the same thing again.

Ali Hatim: And this is what we refused to accept.

I'll tell you something else. Sheikh Sattar Abu Risha, Sheikh Ali—we did not kill anybody with our own hands. It was the fighters from the field who did the killing, yet Sheikh Sattar Abu Risha was made too prominent. He was given much more prominence than he deserved. I'll tell you the truth, and I'm going to speak frankly, he was given too much money under the guise of various projects, meaning arming of the people. But truthfully, just to be frank about it, it was too much money, much more than he deserved. It has to be noted and has to be acknowledged that Sheikh Sattar Abu Risha made a great stand. His stand against al-Qaeda was well known, but not as much as the Americans made it to be. For instance, the fighter in the street, to me, was much more deserving of attention than Abu Risha.

When General Kelly came in, he leveled the field, so to speak. He spoke with everyone equally. There was equality in his treatment. And this is what really we're looking for, this treatment. Do not treat me better than Sheikh Hamid, or Hamid better than Sattar, or Sattar better than so forth and so on—everybody equally.

Montgomery: That prominence is probably what got him killed, though, isn't it?

Ali Hatim: I think you're probably right. The way that his prominence rose may have led to his demise, but Sheikh Sattar Abu Risha had to be fair with his own self, because a man has to be truthful to himself and to the Almighty, too.

When I went to Jordan and tried to get the sheikhs to come back to Iraq to fight, you could see cowardice in their faces: "No, man, leave me alone. I'm not going back."

When the American soldiers go back to the United States, they don't go back there and boast about what they did in Iraq. They simply did their duty, and that is what we did. We did our duty to our country, to our province. The American government provides medical coverage. They pay salaries to their men. In other words, they're taken care of. It's well deserved.

Whereas these sheikhs, when they come back here and say, "Well, look, I am and I am and I am," and they start boasting about their doings, which are false. General Kelly, when he came in, he realized who was who, and he gave each his due. I hope that the new general, who is taking General Kelly's place, will follow in his footsteps.*

What we started in Anbar, we finished in Baghdad. Sheikh Hamid al-Hayess took 24 fighters and went to Hamdiya, and he fought off al-Qaeda. We started many Awakenings in Baghdad. I'll tell you, in one section of Baghdad, the Americans were paying as much as $52 million a month for the Sons of Iraq. In Abu Ghraib, we took part. In southern Baghdad, we took part. Ameriyah, Ghazaliya, also

* MGen Kelly was succeeded as commander of Multi National Force-West by MGen Richard T. Tryon on 9 February 2009.

Adhamiyah. We went forward toward Salah ad-Din, Taji. We used our tribal affiliation and interaction to push this initiative forward. And what was great, when we traveled about in Baghdad, in Salah ad-Din, everywhere else, we found that the American response and cooperation with Iraqis was as good, maybe even better.

I admonished and reminded Obama of the fact that you've accepted the fight in Iraq, so you must pull out responsibly. Do not go out of Iraq as losers because that will be a sad reflection on your legacy, so pull out of Iraq responsibly. Make a responsible pullout, not just because you vowed in your election campaign that you're going to pull out of Iraq—not just to fulfill that vow. No. Make it a responsible pullout, even though, to be frank with you, we are opposed to occupation. We don't want to be occupied, but at the same time, we have to look for a responsible pullout. The most volatile place in Iraq is Anbar Province. But we have to think with our heads and not be emotional.

What we notice right now, the Obama administration is trying to repair things inside the United States. But to be honest with you, the United States proper is affected by what takes place around the world. Obama's emphasis is to go to Afghanistan and do what is right in Afghanistan. Well, let me tell you something: Anbar is much more volatile than Afghanistan. Considering the undercurrent of animosity that exists among various entities, the only reason it is contained right now is because of your presence. It's like a car. You lift your foot off of the accelerator, and the vehicle dies.

I have said what I said as a warning and to be remembered in future years. The security situation right now in Iraq is relevant to many things. Therefore, if we have an irresponsible pullout, it might do more damage, and the situation will become much more dangerous than in 2005.

Interview 8

Sheikh Majed Abd al-Razzaq Ali al-Sulayman

Co-Regent
Dulaimi Tribal Confederation

Sheikh Majed Abd al-Razzaq Ali al-Sulayman is a descendant of Sheikh Ali al-Sulayman, who was one of the most powerful sheikhs in al-Anbar Province during and after the transition from Ottoman to British rule (described in the writings of Gertrude Bell). He is a great uncle of Sheikh Ali Hatim (interview 7 in this book), who is the heir apparent and de facto paramount sheikh of the Dulaimi tribal confederation. Sheikh Majed and his cousin, Sheikh Amer Abd al-Jabbar Ali al-Sulayman al-Assaf, initially served as co-regents due to Sheikh Ali Hatim's youth, with Sheikh Amer residing in Ramadi, Iraq, and Sheikh Majed supporting from Amman, Jordan. Now Ali Hatim has matured, but he continues to honor this protocol out of respect for his uncles.

Sheikh Majed and his family fled to Jordan after coming under government scrutiny following a failed coup against Saddam Hussein in the 1990s. King Hussein of Jordan gave him sanctuary, and he began working with exiled Iraqi opposition groups.

Sheikh Majed was interviewed by Colonel Gary W. Montgomery and Chief Warrant Officer-4 Timothy S. McWilliams on 25 February 2009 in Amman, Jordan.

Sheikh Majed Abd al-Razzaq Ali al-Sulayman: The first one who contacted me, and we are still very close, was Dr. Ayad Allawi. He traveled here to Amman, and I met with him. Two Americans were with him. I don't know who they were. Apparently he informed the Americans in London about Sheikh Majed, and they were astonished and surprised to meet Sheikh Majed here in Amman. I met with Ayad Allawi and the Americans for about two or three hours, and Ayad Allawi was translating for me. They said, "We'll meet you later," and they went back to the States.

A month, two months later, I traveled to London. I met with the opposition in London, [Ayatollah Sayed Muhammad Baqir] al-Hakim—not this Hakim that's here, but the one who was assassinated—Massoud Barzani, and Ahmed Chalabi.

[Conversation with interpreter on what time frame Majed is discussing.]

Majed: That was in the year of 2000. They were talking about the next movement, and where they're going to go, and how they're getting ready to overthrow the regime, and they decided to go and meet in the north of Iraq. I objected to that. I told them, "Look, if I want to enter Iraq, I will not allow myself to get in from the north. If I want to assist, I will enter Iraq from the right road, which is from Jordan to Trebil, and from Trebil to Anbar."

We talked for a long time, and there were so many warnings and objections. You have to be frank, because we're talking about history, and I have to be right in what I'm saying. The only one who spoke who was a patriot to Iraq was Ayad Allawi. He was not sectarian at that time.

I came back to Amman, and we continued our meetings here in Amman. I had a phone call from Dr. Ayad Allawi and Nouri Badran, who was the first interior minister during Ayad Allawi's era. I think that was in 2001 or the end of 2000. He said, "There are highly important VIPs coming from the States, and they want to meet you in person. They want to meet you in Abu Dhabi in the United Arab Emirates."

I traveled to Abu Dhabi, and we met continuously for two or three days. They introduced themselves, but I don't think they gave me their real names. I understood from Nouri Badran that one of them was the assistant to the CIA [Central Intelligence Agency], and the other one was a very important person in the Department of Defense. So we met for two or three days, and they even showed me the plans on how they're going to enter Iraq on a map. This was a long talk. And they told me that a person would be coming from the United States to coordinate with me in person.

At the beginning of 2002, a person came from the States, and I met him here in Amman. His name was George. We used to meet from hotel to hotel for security reasons, and he used to go to the States and come back, but we continued our meetings. He asked me, "What's your influence on the military officers in Anbar Province?". . . I told him, "I'll contact the officers in Anbar in person, right in front of you now." And I said, "I will show you that most of them are not supporters of Saddam Hussein." There were about nine Americans sitting in that meeting. Some of them spoke Arabic. One of the American gentlemen, called Yahir, was translating for me. And I contacted three generals in Anbar. They were listening to what I told them on the other line. One of them was in the Republican Guard, one of them was in military industry, and one was in the artillery. I spoke to them, and they said, "Please, sheikh, assist us in overthrowing this regime today, not tomorrow. We have had enough of Saddam Hussein's regime."

The year 2003 came. We continued meeting during that time. I went to London and back again, and I went over to the United Arab Emirates again. George came back—a very nice gentleman. I think George stayed with me here for two months. We met three or four times a week with his groups, as you know, and we kept looking at Iraqi maps, how to enter Iraq. I showed them how to enter Iraq from the Saudi side, and I even showed them the Iraqi checkpoints. And I told them, "What we need now is to have some people working for us inside Iraq, so they can tell us the Iraqi military movements, the Fedayeen Saddam movements. To check the movements between Abu Ghraib and the border with Jordan, to keep an eye on the borders—the Saudi border, the Jordanian border, and the Syrian border."

The first people we used for formation were the sheep smugglers because nobody keeps an eye on them or thinks of them as spies. They brought their sheep, they kept an eye on Iraqi military movements, and they passed the information to us.

There was a gentleman we met once a week because it was so dangerous at the time to send any written messages. Another gentleman came with George later on, by the name of Roy. George told me that we needed to get some Thuraya satellite phones into Iraq. There were some military guys with them, but they were wearing civilian clothes. At the time, as you know, anyone who had a satellite phone in Iraq would be executed on the spot.

He said, "Let's smuggle 100 satellite phones into Iraq."

I said, "No."

He said, "Let's smuggle 20, 25, up to 30 at a time."

So we used these cell phones between the Syrian border and the Saudi border. They watched the H-3 base and H-2 base, and they kept an eye on the Iraqi checkpoints. They kept passing information every hour.

And we used to meet in hotels for our safety: the Holiday Inn, to the Sheraton, to the Four Seasons—no, the Four Seasons wasn't open—the [Radisson] SAS Hotel. And we used to contact them at 10:00 every night. We used to call everyone, and then we expanded the area. We took it from just the borders to all the way to Ramadi. The first cell phone I handed over to my cousin, Sheikh Amer Abd al-Jabbar, in Ramadi, and that put his life in so much danger. And I handed some to our guys in Fallujah, and some into Abu Ghraib—nearer to the Abu Ghraib base.

Then we had another meeting in London. That was January 2003. I wasn't very happy about that meeting, which was between us, the oppositions. They wanted to go back to the States, and I decided to come back here to Amman. At the meeting, there were the Shi'a Iranians and the Kurds, al-Hakim and Chalabi, Massoud Barzani and Jalal Talabani. They were insisting on dissolving the Iraqi army, and we were objecting to that. We really objected to their idea of dissolving the Iraqi army, and they asked me, "What is the solution?" I said that the Coalition forces enter Iraq, we call it a military coup, and all will be accountable then. The high officers during Saddam Hussein's era, and the politicians, and the Ba'athists, put them aside—the high ones. And the country continues with its organization, with the ministries as they are. Those guys went to the States, and I came back to Jordan.

To be frank, Ahmed Chalabi had a lot of influence in Congress and the Department of Defense. And at State, they talked about the same subject, and Ayad Allawi was objecting to dissolving the Iraqi army. I came back here to Amman, and we continued meeting with our friends. It's a long, lengthy talk. I've got it summarized.

The day that they were warning of invading Iraq from Basrah, the plan changed completely in the last minutes. Roy and George and the other friends took me and some of the opposition from Amman, and they put us on a Jordanian base called H-4, I think, nearer to the Iraqi-Jordanian border. I think 40 of us were there—me, Ayad Allawi, and his groups. Some of the sheikhs from the south were there as well. We stayed there for about 15 or 20 days, I think, maybe 15 days. The war had started by then, and they were telling us how to put on gas masks. I received a phone call from here in Amman. The American embassy wanted me and Dr. Ayad Allawi for an urgent meeting at the American embassy. My people stayed there. We had satellite channels as well at the H-4 base. We were advising the Iraqi military, the Iraqi civilians, "Do not fight the Coalition forces. They are a Coalition of liberation." And we had some Iraqi military officers, generals from the opposition, supporting us as well. One of the generals was Salim al-Imam. He was an ambassador and a soldier. And there was another guy called Ahmad Shabib, who was a military guy, too. And the third one was Ibrahim Janabi. They stayed at the H-4 base, and we came back to Amman.

When I got here to Amman, I was surprised with the news that they attacked Barzan's farm, which was a big disaster to me. The house they attacked was my sister's house, and I lost my sister and her children. I lost 22 members of my family in that attack. My sister and her children and her grandchildren all died in this attack. I was devastated. They hit my old house by mistake, and I think somebody passed wrong information to them. They told them, I think, that this was Barzan's farm and that [Saddam] Hussein was hiding in this house.

I remained here in Amman, and I had my sister waking me up for three days. I received a phone call at 12:00. They said, "Get ready. We are ready to move into Ramadi." Prior to that, the Americans asked me, "What do you think if we start bombing the military bases in Ramadi—in Anbar Province? What do you think of that?"

I told them, "Please do not shoot even one bullet in Anbar Province. I can assure you that once you enter Anbar Province, nobody will resist you, neither the tribes nor the military. And I am responsible for what I say. And anything that happens in Anbar Province, I will be responsible." At that meeting, when I said what I just said, the Iraqi ambassador to Egypt was attending that. His name is Raad al-Alousi. But I advised the Coalition forces, "Do not bomb the province, above all else."

So we left Amman. I think we left Amman about 9:30, 10:00 in the morning, and we were accompanied by our Jordanian friends to protect us. When we got into Iraq, we were very disappointed. I saw that the Trebil border crossing had been looted. There were no employees there. The border was open. There was a big picture of Saddam Hussein at the border. We shot at it with our weapons until it fell on the ground, and we continued into Iraq.

Roy and Ibrahim Janabi entered Iraq before us, and they were waiting for us at the K-90 [highway kilometer marker 90] area. The tribes were waiting for me in my guesthouse, and I was in contact with them all the way by satellite phone, just making sure I was traveling all the way to Ramadi. There were American forces all the way to Ramadi. When I introduced myself to them, they let me go through immediately.

I think we got to Ramadi at around 8:00 in the evening. Ayad Allawi and his group, and Roy and his group, were astonished and amazed at the reception and the welcome that we received from the tribes when we got there. This was recorded by Ayad Allawi on his own camera. There were about 2,000 or 3,000 people from my tribe, and all of them were armed. They were dancing and celebrating. And all the sheikhs were waiting for me, too. We had dinner, and Dr. Ayad Allawi remained in my guesthouse for three or four days. Then he moved into Baghdad after it was secured.

Roy and his group contacted me. They said we needed to have other meetings regarding Anbar Province. I said, "Prior to having our meeting, tomorrow morning I have a meeting with all the tribes of my province. And during our meeting, we'll appoint a governor for the province." The Americans did not believe that we could do that so quickly, and it was the first province in the whole of Iraq. We appointed Abdul Karim Burjis al-Rawi as the first governor for Anbar, and we appointed the first chief of police. His name was Ja'adan [Muhammad al-] Alwani. So what we did, the main thing is that we stabilized the province. We appointed the governor, we appointed the chief of police, and 10 days later, all the government offices reopened, and everybody started going back to work.

So when I went to meet with the Americans, they were very happy about what we had achieved in the province. They called it the first step for democracy and liberation. After 15 days in Anbar, I insisted on reopening and securing the border crossing points by putting sheikhs and Anbaris in charge of the Trebil and Tanf border crossing points.

I remained in Ramadi, and I had several meetings with a general who lived in Saddam Hussein's palace. When we entered the palace, we saw it had been looted completely. The Thurayas had been looted. Thieves looted the whole place. It was a really difficult time that we passed through.

After being in Ramadi 25 days, they requested me for a meeting in Baghdad. Ambassador [Jay M.] Garner was there. All the overseas opposition was there. Some of the opposition from inside Iraq was there, too. That meeting took about three or four hours. "What's next? What's the next step now? What are we going to do?" We said it again, the same as I said it before, "Everything should remain as it was prior to the liberation."

But we faced the biggest tragedy by changing Garner for [L. Paul] Bremer. Bremer came over, and the first thing he did was dissolve the Iraqi army and create an armed enemy against us. This is what we were warning against prior to the liberation. He dissolved all the Ba'athist parties. This is the other disaster he made. He dissolved all the government institutions and ministries. What he did is just like putting a bomb in this room, and the bomb explodes the whole room. He decimated everything.

I wasn't very happy at all then. I was not optimistic. I went out with my American friends to take me back to a meeting with Bremer in a palace in Baghdad. There were 12 of us, me and Adnan Pachachi, Massoud Barzani, Ahmed Chalabi. Ayad was sick. He didn't attend the meeting, but he sent his representative, Nuri Badran. Hoshyar Zebari, the foreign minister, [was there, as was] [Ayatollah Sayed Muhammad Baqir] al-Hakim, the one who was assassinated. We discussed a lot in that meeting regarding the mistakes that he was making by dissolving the Iraqi army and the government institutions, but he insisted on his decisions. The Iranian Shi'as and the Kurds had brainwashed him, and I could not change his mind at all.

The border now became open to everybody who ever wanted to enter in and out of Iraq. Nobody was controlling it. Millions of people entered from Iran. Hundreds of thousands from al-Qaeda and the terrorists entered from Syria.

They wanted to form a governmental committee. Roy contacted me, and the general, and they asked me to come and meet them at Saddam's airport. The road was very dangerous, but they said, "There is an important meeting. You have to attend." They said to enter the palace through the Radwaniyah side. They took me in armored cars. I think there were about six cars. And for history, I have to be straight and frank about what the Americans did. Each American sat on one side, just to secure my life.

I entered the republican palace at the airport, and this was the first time I had entered this palace. All of the government's members, who they wanted to choose, were there. Ghazi al-Yawer was there, as were Pachachi, Allawi, Chalabi, Hakim, Massoud Barzani, and Jalal Talabani. There was seating for groups of people. The Americans, with their arms and their body armor, secured my life, standing next to me, protecting me. And somebody came and asked who I am. It was the British ambassador there. I don't remember his name.

He said to me, "We're going to have a meeting with Bremer now."

And I said, "What for?"

He said, "Once you have the meeting with Bremer, he'll tell you why he wants to talk to you."

It was a lengthy talk with Bremer then, and he was snobby. I told him that I will not be a member of this governing council, and he asked me why. "Because," I said, "you are forming the council wrongly."

And he said, "What am I doing wrong?"

And I told him, "The first mistake you made was dissolving the Iraqi army." I told him then that all the weapons that Saddam Hussein bought for the last 30 years had been looted. I told him, "Go out on the street and see. They're selling weapons on the pavement in streets of Baghdad. This is the first thing." And secondly, now you're starting to form a government on a sectarian basis by choosing from Shi'as, Sunnis, Arabs, and Kurds. I would rather go back and help build my province. That's better than being a member of your government."

So I left. My American friends, who I used to meet here in Jordan, asked me about the meeting. I told them what happened, and they were very sad then.

So I went back, and we made a governing council in Anbar Province. The borders were open. The terrorists started coming in. Every day, there were demonstrations against us. There were no schools at the time. There was no army. There was no Ba'ath Party. There were no security forces. There for two or three months, I was calming people. None of them were receiving salaries, so I was trying my best to keep them quiet.

The governor at the time was an intelligence provider. He said, "We have stores and warehouses full of cement, wood, and metal. I will sell all this stock and get the money for the government before somebody loots it." Then people started looting and stealing what they wanted from the stores. He sold it, and he collected some money. We started paying salaries and calmed the situation.

Then the governing council was formed in Baghdad. I went to Baghdad and used my influence to appoint some Anbaris as guards just to keep them quiet, so they could receive salaries, and the Americans assisted me then, too. We employed about 2,000 or 3,000 Anbaris. The teachers, the lecturers—we managed to get them some salaries, too, with pressure, of course, from the American general.

After that, people started to lose trust. They had enough of people lying to them. They regarded me as if I was the one who contacted President [George W.] Bush and told him to invade Iraq. That's the way people looked on me. The first bomb in Iraq exploded in my guesthouse. As a family, I don't know how God blessed us, and none of us were killed. Some of my guards and people outside were killed. About 25 houses around my guesthouse were demolished completely. The car was a Toyota Land Cruiser, and it was filled with 500 kilograms of TNT. So the American forces came into al-Anbar straightaway and surrounded our area. They were shocked at the amount of TNT that was in this car. With that amount of TNT, they could have demolished half of the city. Anyway, we got out of it safely.

I went to Baghdad for a meeting, me and my bodyguards. I had a license from the general to carry weapons. On the way back from Baghdad (al-Qaeda had taken over Fallujah), I had about 16 or 17 cars in my convoy, and they're all similar, so they didn't know which car I was in. We got to the Saqlawiyah district, and they opened fire on us. Hundreds of thousands of bullets were shooting at us. My guards were brave. Straightaway, they came out of their cars and laid down on the ground. We had PKC [machine guns] with us, and we resisted. It was a valiant resistance. I continued on my way, and my bodyguards stayed and fought. They told me to just leave and go, and there was another car guarding me. I think the Americans heard what was happening. Two Apaches [helicopters] came over, and al-Qaeda ran away. I didn't see, but my bodyguards told me they killed seven al-Qaeda. Of my bodyguards, I think five or six people were injured. Eight of my cars were demolished completely.

We got back to Ramadi, and we continued our work. One day I was calling the governor and coordinating with him. Since my guesthouse was bombed, we were sitting in one of our houses. There was a gentleman trying to get in, and my bodyguard asked him, "Who are you?" He told them, 'I came to see Sheikh Majed, to ask him to appoint me to the police or the army or on the borders." They let him in. It was a shortcoming of my bodyguards. They didn't search him. The important thing is that he entered my house. I was on the phone, speaking to the governor. About 30 of us were sitting there. Sheikh Amer was sitting next to me. My cousin, God bless his soul, Sulayman was sitting there. And Sheikh Ra'ad was by my side as well, and my people were sitting all over.

He asked, "Who is Sheikh Majed." They pointed at me. They just told him in good faith. I asked him straightaway—a boy of 24 or 25 years old—I asked him, "Where are you from?" He said, "I'm a Dulaim," but he said it in a different dialect from Baghdad. I started wondering who this guy is. He started pressing a button. It was winter, and he was wearing a coat. God saved us. He tried to detonate a bomb.

He asked one of the guys who was sitting there, "Is there a toilet around?" We said, "Yes, there is one on the right." My cousins and my nephew said, "There is something wrong with this guy. I think he is full of explosives." Straightaway I told my guards, "Don't let him come in." He went to the toilet. I think the wire wasn't connected to the suicide belt, so he went and connected it. He came running toward me. My bodyguards caught him by both arms. They took him outside the house and threw him into the street. As he fell down, he exploded. Even from a distance, it broke all the glass.

[Rest break. Conversation omitted, except for the following exchange.]

Majed: One thing I would like to mention in your record—you must write it: when I appointed Abdul Karim Burjs al-Rawi as the first governor, I appointed Ahmad Abu Risha as one of his bodyguards—the head of the Awakening.

Interpreter: I said, "Oh, did you appoint Ahmad or Abdul Sattar?" He said, "No, I appointed Ahmad, because Abdul Sattar was a road gangster. He was robbing in the streets."

[end of break]

Colonel Gary W. Montgomery: Could you tell us how the Awakening came about?

Majed: The Awakening—I'm going to speak frankly—the Awakening wasn't built by one person. Unfortunately, credit was given to one man, and from others it was taken. This is the biggest mistake the Americans made, too.

Al-Qaeda started to enter Ramadi city. We don't have it in humanity, not in Islam, not in any other religion, that somebody can come and cut off someone's head. Al-Qaeda started to prohibit people from doing things and to allow other things to be done in the name of Islam, as if they were messengers from God, and it was all lies and wrong. They started to steal.

They did what they did in Fallujah, and I warned the people of Fallujah about them, too, and I told them at the time. I warned the people of Fallujah that al-Qaeda members are not Islamic. They are members of the Iranian groups, and they want to shove us in front of the Americans. This is an Iranian strategy: that we, the Sunnis, are against the Americans and against humanity, and the Shi'as are a peaceful and poor people. Nobody believed me.

When people started to recognize al-Qaeda for what they were, after they destroyed and damaged what they did, it was too late to say, "I'm sorry." They started to realize what I said was right about al-Qaeda, and I told them afterward.

Let's start about the Awakening, how it started. The first thing al-Qaeda did, they started attacking the women. And we, the Arabs, as you know, when you pull a woman's hair, it's a big thing. This can start a war. And there was a woman from the Albu Ali Jassim tribe. She was the daughter of one of the sheikhs. I won't mention names. So the tribes started to rise up after they attacked this woman.

We, all the sheikhs, had a meeting here in Amman. We met to start a revolution, to start the Awakening. We went to [Prime Minister Nouri al-] Maliki and got his advice. He said, "I will not accept the Awakening unless it's blessed by the al-Sulayman tribe." I've got a letter.

I think General [John R.] Allen came to see me at the time, and he asked me, "What's your opinion?" I told him, "What you are doing now, you should have done three years ago."

And, as we say of someone who dies, we just say "God bless him." Sattar was with them, too. Everybody knows the reputation of Sattar, I'm sure you know, and of his brother Ahmad. So we, the sheikhs, discussed among ourselves who will head the Awakening. I tell you frankly—and this is being recorded, so I'm speaking frankly—if any other sheikh denies it, it's on record. They say, "The right job for the right man, and the bad will be accountable unto the bad." So we chose the bad one, which was Sattar, because he was a troublemaker, and he had a bad record, so we chose Sattar. And, as you know, his tribe is a very small tribe.

And in the circumstances that occurred, everyone started to rise up against al-Qaeda. So we sent an e-mail, and I have it here with me. I sent an e-mail to the prime minister, Nouri al-Maliki, advising him that the al-Sulayman tribe blesses this movement, and he can supply them with weapons. And all the Dulaimi tribes signed it, supporting this move. Even General Allen was happy about it.

We held a forum in Ramadi. The whole tribe started supporting it. Each tribesman started to fight against the people within his tribe who were with al-Qaeda, from Fallujah to Rutbah, to Rawah, to al-Hit, to al-Anah, al-Qasiyah, to al-Qaim. Each tribe fought al-Qaeda. This brave work was done by the tribesmen. But unfortunately, it was credited to only one or two people. This is what is not fair.

After we finished, and we cleared al-Qaeda—and thank God we kicked them out—the head of the Awakening, one or two of them, started to abuse it for personal gain, to gain more contracts, only for financial gain.

Now, as you know, the Awakening has been breaking down because al-Qaeda has been kicked out. Each tribe is proud of its achievement, and unfortunately, some of them have been directed into financial gains only. And if you remember, I spoke to General [Martin] Post directly about this matter when he came here. I told him to stop that work. I advised him to stop supporting a few men because it will reflect badly on you. Because the tribes know what is up. They know me, they know you, they know each and every one of us. And they know what you earned and what your financial capability was two or three years ago. Now, after two or three years, I see the difference in them, how much they changed. Now, they have wealth. They're worth millions of dollars. Previously, most of the people who became multimillionaires were road gangsters. They were robbing cars on the highway. So they have been awarded more than they deserve. Now we have to stop the support—quietly and gently, because, as you know, the tribes are very sensitive.

Saddam, with all his power, wanted to play on problems among the sheikhs and open doors for himself. You have to keep this in its history, and keep the tribes to themselves. Each sheikh and each tribe has his history and his position. Saddam made some sheikhs in the '90s, and you know them. And when the Americans came over, they made two or three fake sheikhs, and they're all playboys and road gangsters. This is one of the negative things being done. And, as a friend, I warn my friends in all frankness and honesty.

Montgomery: How can Saddam—or in two or three cases, the Americans—how can they make a sheikh?

Majed: I'll tell you. Saddam opened an office for the tribes, and he started to register the sheikhs: A, B, C. Then, he mixed it up. B became A, C became B. He started mixing the names of the sheikhs as he liked. This is history. You cannot play with history, but he wanted to play with history. The older tradition has its life, has its history.

Now when the Americans came, at the beginning, when we sat with them, they used to call and meet only the main tribal sheikhs in al-Anbar, that is, me and the governor, and the general who used to sit on the panel. And this has all been documented, I'm sure, documented by you. I am the head of all the tribes. [Interpreter—name deleted], for example, is from this tribe, so I bring the head of that tribe. I don't know what Saddam did by changing the names from A to B, B to C and all this, but it does not work with us, because we go back to history, and we know who the head of the tribe is. So in our way, when we met with the Americans, we were on the right path. But once I came to Amman, chaos started—and al-Qaeda. Unfortunately, the Americans don't know our tradition. Everybody who was wearing a scarf like me, they called him sheikh. If he is a sheikh, yes, but any sheikh we meet, go and ask him for his history first.

Montgomery: You mentioned that Sattar was the right man for the right job, and at the same time, you also said that he was from a small tribe. Was it important that he was from a small tribe?

Majed: They are very small. Ask the whole province about the Abu Risha tribe. They are one of the smallest tribes.

Montgomery: What I'm wondering is—

Unknown: Was that part of the criteria for him being chosen?

Interpreter: No, it was the criterion because he was a gangster.

Montgomery: But to start a revolution, wouldn't it be better to have a sheikh from a big tribe?

Majed: Okay. I'll tell you how we chose him. We were hesitant. First of all, we went to Sheikh Tariq, and he said, "Look, I'm a tribal head, and I've got a history. I cannot be a cowboy in the street. It doesn't suit my position."

So we chose someone. He was a son of a tribe. Sattar, God bless his soul. He's dead now. He was a brave man. You have to be fair. And he was not like Ahmad, who loves financial rewards. And he used to talk to everyone. And he educated and civilized himself. He left the past behind, but unfortunately, he was assassinated, and God bless him.

Montgomery: I guess that a small tribe minimizes the risk to other people, doesn't it?

Interpreter: I asked the sheikh, "Was it you chose him because of a small tribe or because he was a gangster?" He said, "That's why, because he was a gangster, and that's why we chose him."

Montgomery: I've heard that some of the inspiration for the Awakening came from a very local uprising in al-Qaim that did not spread anywhere, but it kind of gave an example that people refer back to frequently. What do you think about that?

Majed: I told you before that the Awakening revolution started when al-Qaeda attacked the woman that I mentioned earlier.

Montgomery: And that was in al-Qaim?

Majed: No, it was in Anbar—it was in the Albu Ali Jassim tribe. There was a conflict in al-Qaim, because some of the tribes—if you walk in one path, it's okay, but once you start walking in several paths, you will be discovered. As you know, al-Qaim was the first place al-Qaeda entered. Then they started fighting among themselves. But it didn't last long. It was quieted immediately. But the Awakening started from Ramadi when they took this woman by the hair.

Montgomery: The reason I'm asking these questions is I've interviewed a lot of people, and, as you said earlier, each tribe had its own Awakening, so to speak.

Majed: It's so, it's so.

Montgomery: But when I'm interviewing people, sometimes I hear the same facts from each person, and sometimes I hear things that were more important locally, but not for the whole province. So I ask other people to see if this was a province-wide thing or a local thing.

Majed: I'll tell you why. When you talk to each tribe, it's talking about itself, because each tribe knows who's the infiltrator in the tribe, and they know who's the thief in the tribe. But when I speak, I speak on the whole of the province. The effort we started when we, the sheikhs, met—all of us—in Amman here, we wanted to gather and unite all the efforts of the tribes in one group. And we moved in the right path.

Interview 9

Sheikh Sabah al-Sattam Effan Fahran al-Shurji al-Aziz

Principal Sheikh
Albu Mahal Tribe

Sheikh Sabah al-Sattam Effan Fahran al-Shurji al-Aziz is the principal sheikh of the Albu Mahal tribe in al-Qaim, on the Syrian border. He and his tribe led some of the earliest Iraqi resistance to al-Qaeda in Iraq, in 2005, which is cited by some as one of the first organized "awakening" efforts.

Sheikh Sabah was interviewed by Colonel Gary W. Montgomery and Chief Warrant Officer-4 Timothy S. McWilliams on 25 February 2009 in Amman, Jordan.

Sheikh Sabah al-Sattam Effan Fahran al-Shurji al-Aziz: The western side of Iraq, as we say, is from Haditha to al-Qaim and al-Rutbah. In the beginning, it was invaded by British forces. After two or three months they changed—the American forces took over from the British forces. When they entered as invaders, for sure, it's the right of the people to resist invaders.

One of the biggest mistakes the Coalition forces made at the beginning was getting intelligence information from people that were not trustworthy. And when they got the information, they acted blindly. They started by humiliating the people, scaring children and women. This was the first mistake by the Coalition forces.

Two weeks after, I was detained by the British forces—having done nothing—they detained me on the 17th of April in 2003. It was for false information passed to them by a false man.

We are—we inherited this from our fathers, and grandfathers, and great-grandfathers—we are a brave and traditional tribe in al-Qaim. And we are brave and strong. Some of the tribes—and I say some of them—used to concern some of the smaller tribes, and they wanted to retaliate against us by using the Coalition forces. We, as a tribe, we did not want to create any conflicts among us, especially after the invasion of the Coalition forces. On the contrary, we wanted to keep stability with the Coalition forces and our tribe and other tribes, too, and our traditions.

When the American forces took over from the British forces, unfortunately, the mistakes started to increase. For history and for honesty, the British treatment of us was much better than by the Americans, as the British had more experience with Iraqi people, and in general with the Middle East as a whole. No assassinations were done by the British. It was much less than was done by the Americans, I can say about 70 to 80 percent less killing done by the Coalition than by the Americans.

Unfortunately, the Americans used to deal with violence and toughness. I'm talking about the first two years of the invasion. And they didn't listen to us at all when we told them that the tribes are the keys to solving Iraqis' problems. After the weakening of the tribes, terrorists started entering our province—not only the province, but the whole of Iraq.

We used to meet with the Americans, either through the governor's office or through the local mayor. And we used to tell them frankly in our meetings that the Americans will pay a high price for these mistakes.

It was a disaster when al-Qaeda entered our country, killing and executing Iraqis. Whatever the Tatars did against humanity, al-Qaeda did it worse, worse than anyone that you can think of.

The reason for our conflicts and disagreements with al-Qaeda, there are many points that we disagree on with al-Qaeda. The first thing was al-Qaeda tried to isolate and to humiliate the tribals in the areas. . . . The other thing they did was trying to stop the opinions of the tribal people, to stop the Iraqis themselves from having any opinions on the ground. The only people that we should listen to were them—to al-Qaeda—and these people were not even Iraqis. They were foreigners who entered from Yemen, Saudi Arabia, Syria, and they called themselves emirs. They wanted to have their word and their opinion over us.

In our personal experiences with them, they dealt violently and toughly with all Iraqi citizens. They accused every Iraqi of being a traitor or a spy for the Americans. And if you weren't a spy for the Americans, then you should listen to them and kill Iraqis, and execute them, and steal money from Iraqis, and then they can think you are better to them than to the Americans.

The other point is that they chose to have supporters in our areas. They entered from the gate of martyrdom. They said, "We came to help you fight the invaders who invaded Iraq." In this matter, they used two paths. The first path was by scaring people, and the other path was financial. The families were in financial difficulties. None of them had any finances left on them. They were spending huge amounts of money, giving to the people. So in this way, they gathered supporters around them. And if anyone in the area said that al-Qaeda were not Muslims, nobody supported him or believed what he said.

People like us, and others who were quite wise and educated and professional, started to realize who these people were. Some of them were from the former army. Some of them were notable people who had the wrong vision. They started to know that these came in, abusing the religion of Islam, and they were abusing their people by executing innocent people. People started to realize that they were blackening the word "Muslim," so they realized what their mission was.

Then, at the end of 2004, the conflicts between us and them started. And we continued these conflicts on a low level up to May 2005.

Colonel Gary W. Montgomery: That began in 2004?

Interpreter: That began in 2004, when the conflict started, the low-level conflicts, and then up to May 2005.

Sabah: When you start to find the bold lines among the lines, we realized what their mission was, we in al-Qaim, especially our tribe, the Albu Mahal—and the whole of Iraq. We started to understand their mission was to destroy our tribe.

The first thing they did, they assassinated the chief of police, who was a member of our tribe. And then I said, "That's it. Now the time has come." And we started fighting them, and we removed them from the center of al-Qaim.

Unfortunately, we didn't have the arms to fight them at that time because the Coalition forces were surrounding us. Despite the American forces around us, we managed to get rid of al-Qaeda inside our cities, our villages. But militarily, al-Qaeda was even more in control after we kicked them out, because they started to surround the area. They sent a representative to us to have a dialogue so they could reenter al-Qaim. We rejected that.

The big conflict was in August 2005 after they arrested the governor, who was one of our cousins. Unfortunately they assassinated him later. And then after they assassinated our cousin, we said, "That's it. We have to fight against them." I called the minister of interior, Mr. Sadoon al-Dulaimi, and he was here in Amman.

I want to go back to what I said. In August 2005 and the fighting with al-Qaeda, I wanted to mention to you that al-Qaeda won. They reentered al-Qaim, and we had to withdraw. I think the number who entered al-Qaim were about 4,000 to 5,000 al-Qaeda members. And this is where the Americans made a mistake again. These al-Qaeda members who came all this distance—they came from Bayji and Mosul all the way to al-Qaim, and the Americans couldn't even stop them. They used to see them, but the Coalition forces didn't do anything. I think this was a big mistake by the Coalition forces in the human rights of the women and children who lost their lives.

So what I did, we built a military division with the minister of interior, and we called it the Desert Protectors—less than 400 people, 98 percent of them from my tribe. With God's blessing and the support of a small number of the Iraqi army, we managed to prevent al-Qaeda from entering al-Qaim again. And thank God, things got better from that time until now. We cleaned it up.

And then after a year, a year and a few months, the Ramadi Awakening started. A number of people speak about the Awakening in Ramadi. I'm not a member of the Ramadi Awakening. And another Awakening started in Haditha prior to Ramadi. But there were only a small number in Haditha. And then the whole province united, and the Awakening started to move to Mosul and Salah ad-Din and Baghdad and Diyala. And Iraq, as you see it, has been stabilized, and we thank the Lord for his assistance and the tribal leaders. Just to emphasize without going back again, as I said earlier, the tribes are the key to stabilization and security in Iraq.

Montgomery: What role did the Americans play in the Awakening, if any?

Sabah: In the Ramadi Awakening, the Americans played a big role. It was the American support for the Awakening that made it really successful, because they are the main force on the ground. And then we felt that the Americans started to change their behavior, and they were working in the right way. But they did play a big role, I have to admit.

Chief Warrant Officer-4 Timothy S. McWilliams: What year did they change their behaviors?

Sabah: In 2006.

Montgomery: Which officers do you remember playing a role in the Awakening?

Sabah: I believe General [John R.] Allen.

Montgomery: What effect did the two battles in Fallujah have on Anbar Province as a whole?

Sabah: In the first conflict, there was support for the resistance. In the second Fallujah, it changed the balances. In the first, as I said, many people sympathized with the resistance, and the second one changed the view and the vision of the people against al-Qaeda, because they started to realize who al-Qaeda are. Al-Qaeda are people who kill, demolish houses, rape people, so the people started to change their view of the resistance.

Montgomery: Is that because of the behavior of al-Qaeda in Fallujah in between the battles?

Sabah: Yes. It was in the whole of the province, not just Fallujah.

Montgomery: Okay. Did the battle of Sufiyah have a larger effect on the rest of the province? Are you familiar with it?

Sabah: As you know, I'm far from Ramadi. I'm 300 kilometers away from Ramadi. The same thing that happened, the conflicts happened in al-Qaim, happened in Haditha and Ramadi.

Montgomery: Right. We've heard about al-Qaeda grabbing a woman by the hair and pulling her out of a car. Are you familiar with such an incident?

Sabah: They did worse things than that. I'll give you an example of one of the things they did. Four members of one family in the al-Qaim district—not from my tribe, from the Salman tribe—they were former soldiers. They brought them out and killed them in the middle of the street. They cut off their heads, and they put each head on the back of the body. This was in August of 2004. The family came to pick them up, to take them and bury them. Al-Qaeda stopped them. I contacted the director general for the phosphate factory to bring a shovel so he can just bury the bodies. It was starting to smell. They stopped them. After a week or 10 days, if a car wanted to pass through this area, he had to put a mask on his face from the smell. What they did, they came and put TNT on each body, and they called the family to come pick them up. When the family came at dawn, they exploded them. With these four people, they killed another eight, and they injured many others.

I will ask a question. If you are a Muslim, a Christian, Jewish, Buddhist, any religion, for what reason do you allow something like this to happen? "I am a martyr" is the excuse that al-Qaeda used. This is one of the hundreds of thousands of cases that happened. This is what alienated people from the real resistance. This is what I've seen from my own eyes. I've seen the bodies. And we tried to bury them, but we couldn't.

For example, I went to Baghdad. From al-Qaim to Baghdad is 400 kilometers. In these 400 kilometers, I would see about 10 to 20 bodies lying on the street beheaded. So for them to pull a woman's hair is nothing compared to what they did to the other bodies.

A man was accused of killing four Shi'as in the al-Qaim district. I said to him, "You are accused of killing these four people." His father was with him. He said, "No, I didn't kill anyone. But if the

Saudi emir tells me that I killed them, I will say 'Yes, I killed them.'" This is the mentality that I'm trying to explain to you, that al-Qaeda were brainwashing the people. The same man told me, "If the al-Qaeda emir told me to assassinate my father, I will assassinate him." They were brainwashed to that extent.

We went through hard times, very, very hard times.

Montgomery: Where do you think their funding came from? You said they spent a lot of money.

Sabah: Iran. Iran was the number one in supplying the financial aid. And I can say tens of millions of dollars were supporting al-Qaeda. Imagine a boy of 15 or 16 years old. Imagine he's got five or six cars outside his house. Where would he get this money?

Montgomery: How is it that if these men were from Yemen, Saudi Arabia, and other places, which is on the Sunni side of things. Why would they work with Iran?

Sabah: He came to assassinate the Sunnis because he's being paid to kill the Sunnis.

Montgomery: Even though he's from a Sunni country?

Sabah: Yes.

Montgomery: Would it be accurate to say that they could make that arrangement because religion was not involved?

Sabah: This is political and traditional in what they believe. They will agree to assassinate and kill anyone. He will change his religion and his tradition and his belief for the sake of money.

Montgomery: So this was really political and criminal.

Sabah: Yes, exactly. . . .

Of course after the Awakening formed, the most important thing is that the killing had stopped in the province. In one or two months, you see nobody has been killed, while prior to that, five people were being killed on a daily basis. That's why it goes back to what I'm saying, that the Americans put the tribes aside, and what the right thing is, is that the tribes did play a role. This is the most important thing, that civilians can walk on the street safely.

The mistake that has been made on the national level is that you brought people that are not Iraqis into the central government, and up to this minute, most of them are not loyal to Iraq. And this is the worst, if the Americans withdraw.

I think if the Americans withdraw, two things will happen. Iran will enter through the militias. That is one. This is a big danger to us. The second one, the Kurds will start all over again, the violence will start.

My advice to the Americans: we are not a democratic nation, and democracy does not suit us at present. It's going to take 10, 15 to 20 years for the people to understand the meaning of democracy. I tell you this as advice for history. The Sunnis have to play a role, their natural role, and all of these things about elections, and this and that, is all forgery. Take the election in Anbar two weeks ago, where most of them are Sunnis. There was forgery and corruption. So what I can advise is that we are not ready for democracy yet. People are not ready for it. We have to be fair with the Sunnis. And I believe that for the benefit of America that the one who rules Iraq should be Sunni, and he will protect Americans' interest.

We have a saying that "you cannot hold two pomegranates in one hand." One of the pomegranates is America, which is big. Just a little bit smaller than the other pomegranate is Iran. So any Shi'a rulers who come to Iraq—especially the Islamic ones—he is linked directly in his beliefs to Iran. He can never serve American and Iranian interests at the same time, which are in opposition to each other. So what we say, that a man who comes, it could be a secular gentleman, who could be a Sunni or a Shiite. This is an exit for the Iraqis' crisis. This is my opinion. We believe this is the right solution for Iraq.

Montgomery: This is for clarification. You said the Americans brought non-Iraqis into the government. Are you referring to Iraqi Shi'as and Kurds, or are you referring to Iraqis who are in the pay of Iran, or something else?

Sabah: Their loyalty. Their loyalty is not to their country. [Abdul Aziz] al-Hakim, in one of his speeches—and he was the head of the governing council—asked Iraq to pay compensation to Iran. At the same time, many of the international countries were writing off Iraqi debts. So his loyalty is not to Iraq. Look at their militias and the killing they've been doing. We, as the Sunnis, we fought al-Qaeda, and we eliminated al-Qaeda. As you know, at the Ministry of Interior, there are people who speak Iranian, and other locations, too. Iranian intelligence—I don't know what they call it—they are the ones who are ruling the country. The Iranian infiltration into Iraq is very strong.

So for the present government, preparing themselves for the new coming election, they are just speaking with the Sunnis for the time being, just wasting their time. And if this present government is reelected again, you will see for yourself that Iranian influence and Iranian power will increase more than it is now. Even the Sunnis who are in the government now do not represent all the Sunnis.

So my advice to the Americans: they should know how to deal with the Sunnis so they can gather maximum support from the Sunnis. And I hope many matters can be changed, because Iraq has sacrificed a lot, economically and in blood.

Look at Jordan. Jordan's a respectable country. Two days ago, I came from Egypt. It's a nice country. Why in Iraq, up to this moment, is there no development? We have no roads. People are talking about security, security, security. Why? It's been six years since Iraq was invaded. There's no electricity, no clean drinking water, and to my knowledge, Iraq is the richest country in the world. We have gas, oil, agriculture, and professional people with great minds. We know the answer. The answer is the people who are responsible for ruling the country are not Iraqis. We were wishing, with the oil prices going up so high, we were expecting some construction and infrastructure stuff to start to erupt. Billions of dollars have been spent from the American treasury and the Iraqi treasury. The money they spent could develop two countries, not one. Where is the development? Where is it? Nothing. You don't see anything.

Political
Perspectives

جمهورية العراق
مجلس محافظة الانبار
وامرهم ... بينهم

Interview 10

Governor Mamoun Sami Rashid al-Alwani

Governor
al-Anbar Province

Governor Mamoun Sami Rashid al-Alwani is a member of the Iraqi Islamic Party. Before the 2003 invasion of Iraq, he worked as an engineer in the housing and construction industry. In 2004, he ran for political office and was elected to the Anbar provincial council, which chose him as its chairman at the beginning of 2005. In June 2005, the provincial council selected him to serve as governor of the province.

In accepting the position, he would have been acutely aware that his predecessor, Governor Raja Nawaf Farhan Mahalawi, was kidnapped and killed. Also, he would have known that Governor Abdul Karim Burghis al-Rawi had resigned in 2004 after his sons were kidnapped and his house set ablaze.

Governor Mamoun was protected by Marines and his own personal security detail. Nevertheless, he was the target of 45 assassination attempts during his term as governor, and his 12-year-old son was kidnapped in September 2005. Governor Mamoun took an advance on his salary, paid a $6,000 ransom, and his son was returned the next day. He continued to work.

On 1 September 2008, he had the satisfaction of representing al-Anbar in the Provincial Iraqi Control ceremony, which transferred security responsibilities in al-Anbar province from American to Iraqi control.

Governor Mamoun was interviewed by Colonel Gary W. Montgomery, Lieutenant Colonel Bradley E. Weisz, and Chief Warrant Officer-4 Timothy S. McWilliams on 16 February 2009 at the Provincial Government Center in Ramadi.

Governor Mamoun Sami Rashid al-Alwani: The problem with Anbar is that it's a tribal-oriented society. The tribes are a bunch of groups that are attached to one entity, and these tribes are spread throughout the districts here in Anbar, which is made up of eight districts.

After the fall of the [Saddam Hussein] regime, there was a big vacuum government-wise. The province did not get into a struggle and did not play a role in the fall of the regime. The Coalition forces and the government that was in place here kept the government working normally, so to speak. The tribal leaders were very careful to ensure that they kept the province secure and stable and protected the people. The quiet period lasted until April 2004, maybe even until the end of 2004.

At that time, an ideology developed that went against the security of the province. That ideology included religious extremism, sectarianism, which was just being established. There were military operations, starting in Fallujah. In 2004, there were great battles that caused the destruction of the city and caused many people to leave the city. There were also, at the same time, people who were trying to establish security and establish the rule of law, to establish a mayor, a council, specifically in Fallujah.

Unfortunately—because we did not have conscientious people, and because of speeches given by imams that misled the people—they created an internal struggle. That caused attacks against the Coalition forces, and of course they started challenging the legal system. It required a lot of sacrifices from many, many people, but in the end, the biggest loss was the city itself. Terrorism found a door by taking advantage of uneducated people and implanting extremist religious ideology, supported by money.

There was also administrative or governmental breakdown. One of the biggest breakdowns was the open borders that we had with other countries, which gave access to people to come into Iraq. Many foreigners spread out in the province. We felt, as soon as that happened, that these people were not actually resisting the presence of the Americans inside the country, but they came to politicize their own agendas on the people. Many people were drawn into their ideology, and we felt it tangibly.

Made in the USA
Monee, IL
06 November 2025

33942615R00095